CRYSTAL SYMMETRY:
Theory of Colour Crystallography

ELLIS HORWOOD SERIES IN
MATHEMATICS AND ITS APPLICATIONS
Series Editor: Professor G. M. BELL, Chelsea College, University of London
(and within the same series)
Statistics and Operational Research
Editor: B. W. CONOLLY, Chelsea College, University of London

Baldock, G. R. & Bridgeman, T.	MATHEMATICAL THEORY OF WAVE MOTION
Beaumont, G. P.	INTRODUCTORY APPLIED PROBABILITY
Burghes, D. N. & Wood, A. D.	MATHEMATICAL MODELS IN SOCIAL MANAGEMENT AND LIFE SCIENCES
Burghes, D. N.	MODERN INTRODUCTION TO CLASSICAL MECHANICS AND CONTROL
Burghes, D. N. & Graham, A.	CONTROL AND OPTIMAL CONTROL
Burghes, D. N., Huntley, I. & McDonald, J.	APPLYING MATHEMATICS
Butkovskiy, A. G.	GREEN'S FUNCTIONS AND TRANSFER FUNCTIONS HANDBOOK
Butkovskiy, A. G.	STRUCTURE OF DISTRIBUTED SYSTEMS
Chorlton, F.	TEXTBOOK OF DYNAMICS
Chorlton, F.	VECTOR AND TENSOR METHODS
Conolly, B.	TECHNIQUES IN OPERATIONAL RESEARCH Vol. 1: QUEUEING SYSTEMS Vol. 2: MODELS, SEARCH, RANDOMIZATION
Dunning-Davies, J.	MATHEMATICAL METHODS FOR MATHEMATICS, PHYSICAL SCIENCE AND ENGINEERING
Eason, G., Coles, C. W., Gettinby, G.	MATHEMATICS FOR THE BIOSCIENCES
Exton, H.	HANDBOOK OF HYPERGEOMETRIC INTEGRALS
Exton, H.	MULTIPLE HYPERGEOMETRIC FUNCTIONS
Faux, I. D. & Pratt, M. J.	COMPUTATIONAL GEOMETRY FOR DESIGN AND MANUFACTURE
Goult, R. J.	APPLIED LINEAR ALGEBRA
Graham, A.	KRONECKER PRODUCTS AND MATRIX CALCULUS: WITH APPLICATIONS
Graham, A.	MATRIX THEORY AND APPLICATIONS FOR ENGINEERS AND MATHEMATICIANS
Griffel, D. H.	APPLIED FUNCTIONAL ANALYSIS
Hoskins, R. F.	GENERALISED FUNCTIONS
Hunter, S. C.	MECHANICS OF CONTINUOUS MEDIA, 2nd (Revised) Edition
Huntley, I. & Johnson, R. M.	LINEAR AND NON-LINEAR DIFFERENTIAL EQUATIONS
Jones, A. J.	GAME THEORY
Kemp, K. W.	COMPUTATIONAL STATISTICS
Kosinski, W.	FIELD SINGULARITIES AND WAVE ANALYSIS IN CONTINUUM MECHANICS
Marichev, O. I.	INTEGRALS OF HIGHER TRANSCENDENTAL FUNCTIONS
Meek, B. L. & Fairthorne, S.	USING COMPUTERS
Muller-Pfeiffer, E.	SPECTRAL THEORY OF ORDINARY DIFFERENTIAL OPERATORS
Nonweiler, T. R. F.	COMPUTATIONAL MATHEMATICS: An Introduction to Numerical Analysis
Oliviera-Pinto, F.	SIMULATION CONCEPTS IN MATHEMATICAL MODELLING
Oliviera-Pinto, F. & Conolly, B. W.	APPLICABLE MATHEMATICS OF NON-PHYSICAL PHENOMENA
Scorer, R. S.	ENVIRONMENTAL AERODYNAMICS
Smith, D. K.	NETWORK OPTIMISATION PRACTICE: A Computational Guide
Stoodley, K. D. C., Lewis, T. & Stainton, C. L. S.	APPLIED STATISTICAL TECHNIQUES
Sweet, M. V.	ALGEBRA, GEOMETRY AND TRIGONOMETRY FOR SCIENCE STUDENTS
Temperley, H. N. V. & Trevena, D. H.	LIQUIDS AND THEIR PROPERTIES
Temperley, H. N. V.	GRAPH THEORY AND APPLICATIONS
Twizell, E. H.	COMPUTATIONAL METHODS FOR PARTIAL DIFFERENTIAL EQUATIONS IN BIOMEDICINE
Whitehead, J. R.	THE DESIGN AND ANALYSIS OF SEQUENTIAL CLINICAL TRIALS

CRYSTAL SYMMETRY:
Theory of
Colour Crystallography

M. A. JASWON, B.Sc., M.A., Ph.D.
Professor and Head of Department of Mathematics
The City University
London

and

M. A. ROSE, B.Sc., Ph.D.
Head of Computer Business Studies
Henry Thornton School
London

ELLIS HORWOOD LIMITED
Publishers · Chichester

Halsted Press: a division of
JOHN WILEY & SONS
New York · Brisbane · Chichester · Toronto

First published in 1983 by
ELLIS HORWOOD LIMITED
Market Cross House, Cooper Street, Chichester, West Sussex, PO19 1EB, England

The publisher's colophon is reproduced from James Gillison's drawing of the ancient Market Cross, Chichester.

Distributors:

Australia, New Zealand, South-east Asia:
Jacaranda-Wiley Ltd., Jacaranda Press,
JOHN WILEY & SONS INC.,
G.P.O. Box 859, Brisbane, Queensland 40001, Australia

Canada:
JOHN WILEY & SONS CANADA LIMITED
22 Worcester Road, Rexdale, Ontario, Canada.

Europe, Africa:
JOHN WILEY & SONS LIMITED
Baffins Lane, Chichester, West Sussex, England.

North and South America and the rest of the world:
Halsted Press: a division of
JOHN WILEY & SONS
605 Third Avenue, New York, N.Y. 10016, U.S.A.

©1983 M. A. Jaswon and M. A. Rose/Ellis Horwood Ltd.

British Library Cataloguing in Publication Data
Jaswon, M. A.
Crystal symmetry. —
(Ellis Horwood series in mathematics and its applications).
1. Symmetry (Physics) 2 Crystallography
I. Title II. Rose, M. A.
548'.81 QC921

Library of Congress Card No. 82-21380

ISBN 0-85312-229-6 (Ellis Horwood Ltd., Publishers — Library Edn.)
ISBN 0-85312-520-1 (Ellis Horwood Ltd., Publishers — Student Edn.)
ISBN 0-470-27353-4 (Halsted Press — Library Edn.)
ISBN 0-470-27379-8 (Halsted Press — Student Edn.)

Printed in Great Britain by Unwin Brothers of Woking

Contents

Preface..9
Introduction...11
List of Symbols..13

Part I: Crystallographic Point Groups 15

1   **Symmetry Patterns**
    1.1 Introduction 15
    1.2 Symmetry operations..................................15
    1.3 Isolation of principal symmetry axis..............18
    1.4 Combination of symmetry elements 19
    1.5 Inversion centre 23

2   **Mathematical Formulation**
    2.1 Crystallographic point groups 26
    2.2 Cyclic groups 27
    2.3 Abelian groups.......................................28
    2.4 Dihedral groups 31
    2.5 Subgroup structure 34
    2.6 Conclusions...35

3   **Cubic Symmetries**
    3.1 Introduction 37
    3.2 The tetrahedral group 38
    3.3 The octahedral group 40
    3.4 Symmetry of regular tetrahedron and cube 41
    3.5 Inversion effects 44
    3.6 Degeneracies...47

4   **Colour Point Groups**
    4.1 Colour symmetry 48
    4.2 Cyclic colour groups 49
    4.3 Dihedral colour groups.............................50
    4.4 Abelian colour groups...............................55

4.5 Extended dihedral colour groups 55
4.6 Cubic colour groups 58
4.7 Conclusions...59

Part II: Space Lattices 61

5 Lattice Geometry
5.1 Two-dimensional lattices 61
5.2 Three-dimensional lattices 64
5.3 Unit cells...67

6 The Seven Crystal Systems
6.1 Rotational symmetry of two-dimensional lattices............68
6.2 Two-dimensional crystal models 70
6.3 The seven crystal systems 73
6.4 Three-dimensional crystal models 76
6.5 Close-packed hexagonal structure 76

7 Non-primitive Unit Cells
7.1 Body-centred cells 80
7.2 Face-centred cells......................................81
7.3 End-centred cells 83
7.4 Rhombohedral cells.....................................84
7.5 Symmetry of non-primitive Bravais cells 86

8 Translation Groups
8.1 Translation operators 88
8.2 Coupling between $\{\mathscr{T}\}$ and $\{G\}$ 89
8.3 Coloured translation operators 90
8.4 Edge-centring 94

Part III: Space Groups 98

9 Symmorphic (Bravais) Space Groups
9.1 Introduction 98
9.2 Classification of symmorphic space groups 100
9.3 Ambiguities of setting.................................100

10 Screw Axes
10.1 Macroscopic and microscopic symmetry 104
10.2 Cyclic groupoids 105
10.3 Significance of $\{n_1\} \otimes \{\mathscr{T}\} = \{\mathscr{T}\} \otimes \{n_1\}$.....................108

11   Principal and Secondary Screw Axes
   11.1 Theory of composite operators 110
   11.2 Dihedral groupoids: principal screw axes 111
   11.3 Dihedral groupoids: secondary screw axes 113
   11.4 Cubic groupoids.......................................116

12   Glide Planes
   12.1 Glide operators 119
   12.2 Abelian glide groupoids 120
   12 3 Dihedral glide groupoids 121
   12.4 Further orthorhombic groupoids 125
   12.5 Further tetragonal groupoids...........................127
   12.6 Further hexagonal groupoids...........................128
   12.7 Further cubic groupoids 129

13   Diamond Glide
   13.1 Space groups: centred lattices 133
   13.2 Equivalent space groups 134
   13.3 The space group $\left\{\dfrac{4_1}{a}\right\}\otimes\{\mathcal{T}\}$.....................137
   13.4 Diamond glide 138
   13.5 Cubic systems..140
   13.6 Summary...143

14   The Colour Space Groups
   14.1 Colour space groups: Bravais type 144
   14.2 Colour screw groupoids 149
   14.3 Colour glide groupoids: monoclinic and orthorhombic
       systems..152
   14.4 Colour glide groupoids: tetragonal system 156
   14.5 Colour glide groupoids: hexagonal system 157
   14.6 Colour glide groupoids: cubic system 158
   14.7 Summary...159

Appendixes
   1a   Choice of Co-ordinate System...........................161
   1b   Representation of Rotation Operators 164
   2   Combination of Crystallographic Axes 166
   3   Some Elementary Group Theory 170
   4   Class Structure of Point Groups 171
   5   Packing and Stacking of Lattice Planes 172
   6   Composite Operators 175
   7   Two-dimensional Space Groups 178
   8   Two-dimensional Colour Space Groups 181
   9   Names and Symbols for the Crystal Classes 184

References 186

Preface

Most expositions of crystal symmetry emphasise either a descriptive approach or a mathematical approach to the subject. The descriptive approach provides insight into the geometrical features of symmetry relationships, which is particularly valuable for students of crystal physics, materials science, metallurgy and mineralogy. The mathematical approach provides a more fundamental treatment necessary for professional crystallographers, theoretical physicists and applied mathematicians. Each remains incomplete without the other, and the increasing interest in colour symmetry has sharpened the need for an integrated approach. This book builds upon an earlier text by one of us [1] dealing with the classical space groups, and upon subsequent work by the other [2] dealing with the colour space groups (also termed magnetic, black-and-white, or Shubnikov groups). It is offered as a new, distinctive, contribution to the subject. The text is supported by numerous diagrams, including some unique coloured stereograms, and by many tables including some fresh breakdowns of available information.

Acknowledgements are due to Bradley & Cracknell [3], and to Shubnikov & Koptsick [4] for their full accounts of the colour space groups; to Phillips [5] and Buerger [6] for basic crystallographic information; and to Wilson [7] and Woolfson [8] for their examples of applications. Fig. A7.1 has been essentially reproduced from Bhagavantam & Venkatarayudu [9] with their publisher's permission.

Thanks are due to Dr. T. E. Stanley, Department of Mathematics, The City University for helpful discussions on Group Theory; to Prof. E. G. Steward, Department of Physics, The City University, for useful crystallographic information; and to Mary Williams, Department of Mathematics, The City University, for her careful typing. Finally, thanks are due to Ellis Horwood for his patience and ready co-operation at all times.

June 1982
M. A. Jaswon
M. A. Rose

Introduction

This book divides naturally into three main parts. Part I provides a fresh account of the classical point groups and of the colour point groups. We start with intuitive geometric considerations which are then supplemented by formal group theory. The slight knowledge required of the latter topic may be found in numerous texts, e.g. Bhagavantam & Venkatarayudu [9], Lederman [10] and will therefore not be expounded here. Part II covers space lattices and their symmetry properties, including the extension to colour space lattices. Also, some simple crystal models are discussed in order to fix ideas and pave the way for the general theory. Part III formulates space-group theory with particular emphasis on the motif pattern, i.e. the smallest microscopic arrangement of atoms which generates the whole crystal structure by translational repetition. This was pictured by Bravais [11] as a miniature replica of the macroscopic crystal which it produces, implying that its symmetry conforms to one of the crystallographic point groups. Although microscopic symmetry elements were later extended by the introduction of screw axes and glide planes, the Bravais picture nevertheless provides a useful starting point into the theory, apart from directly accounting for nearly one-third of the classical space groups and more than one-fifth of the colour space groups.

The central problem of mathematical crystallography is to determine the independent microscopic symmetries consistent with every macroscopic crystal symmetry. For instance, given that a crystal conforms to the symmetry 6 (6-fold axis of rotational symmetry) on the basis of physical and goniometric data, we may infer that its underlying space lattice must be primitive hexagonal (P). However, the motif pattern may conform to any of the symmetries $6_1, 6_2, 6_3, 6_4, 6_5$ (i.e. screw variants of 6) as well as to 6 itself, so providing the space-group possibilities $P6_1, P6_2, \ldots, P6_5$ in addition to the Bravais space group $P6$. A more complicated example occurs with the macroscopic crystal symmetry 3 (3-fold axis of rotational symmetry). Here the underlying space lattice may be either primitive (P) or doubly-centred hexagonal (i.e. rhombohedral, symbolised (R), so providing the space-group possibilities $P3, P3_1, P3_2$; $R3, R3_1, R3_2$. However the latter three prove to be equivalent, yielding only the single independent space group $R3$.

We have introduced the term 'groupoid' to denote the set of space-group operators which generate all the equivalent atoms of a motif pattern from any one of them. The groupoid reduces to a point group—the isogonal point group—when its screw and glide operators are replaced by the corresponding pure rotation and reflection operators. The central problem, alternatively expressed, is to construct all the independent groupoids isogonal to every point group. It is a straightforward step to expand a groupoid into admissible space groups, but these often turn out to be equivalent as exemplified above. We are not aware of any easy algorithm for coping with the equivalence problem in all the cases which arise.

A fresh symbolism has been utilised for point-group operators, which is readily extended to screw and glide operators. This enables us to display every classical and coloured groupoid in a form which exposes its main geometrical features. The equivalent atomic positions may be quickly computed relative to a naturally available co-ordinate system, i.e. that defined by the principal symmetry elements of the isogonal point group. Our co-ordinates do not always coincide with those written in the International Tables [12], since the latter may refer to a different choice of origin (motivated by practical considerations of structure determination) and possibly a different identification of the co-ordinate axes. Generally speaking, the language of Buerger [6] has been followed, apart from some minor changes as befits a work of purely theoretical orientation.

Expositions of the theory vary in emphasis from the early geometrical arguments of Hilton [13] to the more recent algebraic apparatus of Schwarzenberger [14]. An individualistic analysis of two-dimensional colour patterns has been given by Loeb [15]. The theory is now essentially complete and no new results can be expected. However, it will continue to fascinate by virtue of its mathematical perfection and physical relevance, always awaiting the challenge of fresh presentations and formulations.

List of Symbols

I

A, B	rotations through π about [100], [010]
$\mathscr{A}, \mathscr{B}, \mathscr{C}$	lattice translation operators
A, B, C	end-centred cells
C	rotation through $2\pi/n$ about [001]
D	reflection through axial mirror plane or rotation through π about a secondary axis
$\{\mathscr{E}\}$	end-centred translation group
F	face-centred cell
$\{\mathscr{F}\}$	face-centred translation group
G_i	point-group operator
$\{G\}$	point group
$\{G'\}$	colour point group corresponding with $\{G\}$
Γ_i	screw or glide operator corresponding with G_i
$\{\Gamma\}$	groupoid corresponding with $\{G\}$
H	subgroup of $\{G\}$ of index 2
I	identity operator
I	body-centred cell
$\{\mathscr{I}\}$	body-centred translation group
J	inversion operator
M	reflection through transverse mirror plane
P	primitive cell
R	primitive rhombohedral or doubly-centred hexagonal cell
$\{\mathscr{R}\}$	doubly-centred translation group
R	arbitrary position vector
S	colour switch operator
$\mathscr{T}$	translation operator of form $\mathscr{A}^\lambda \mathscr{B}^\mu \mathscr{C}^\nu$
$\{\mathscr{T}\}$	infinite translation group

II

a, b, c	generating vectors
a, b, c	symbols for axial glide
d	symbol for diamond glide
$[hkl]$	symmetry axis having direction ratios $h:k:l$

$[hkl]^{1/n}$	fractional translation operator in direction $[hkl]$
(hkl)	Miller indices of lattice plane
$(hkl)_m$	reflection through lattice plane (hkl)
$(hkl)^{1/n}$	rotation through $2\pi/n$ about $[hkl]$
$\lambda,\ \mu,\ \nu$	rational numbers
$\left(\frac{\lambda}{2}\ \frac{\mu}{2}\ \frac{\nu}{2}\Big/hkl\right)^{1/n}_{+} = \mathscr{A}^{\lambda}\mathscr{B}^{\mu}\mathscr{C}^{\nu}\ (hkl)^{1/n}$ i.e. a rotation followed by a transverse translation providing a pure rotation	
m	symbol for mirror reflection
g	symbol for diagonal glide
$\{n\}$	cyclic group of order n
$\{n'\}$	colour cyclic group isomorphic with $\{n\}$
$\{n_i\}$	groupoid corresponding with $\{n\}$
$\{n'_i\}$	colour groupoid corresponding with $\{n_i\}$
$\mathbf{t}$	arbitrary lattice vector
$[x,\ y,\ z]$	lattice point of co-ordinates $x,\ y,\ z$
$<x,\ y,\ z>$	lattice vector with components $x,\ y,\ z$

III

$=$	geometrical and mathematical equivalence
$\equiv$	same but in a different geometrical orientation
$\Rightarrow$	equality up to an appropriate member of $\{\mathscr{T}\}$
$\Leftrightarrow$	isomorphic (but not geometrical) equivalence
$\rightarrow$	homomorphic equivalence
$\subset$	inclusion symbol
$\not\subset$	exclusion symbol

PART I: CRYSTALLOGRAPHIC POINT GROUPS

Chapter 1

Symmetry Patterns

1.1 INTRODUCTION

Crystallographic symmetry has two aspects, the macroscopic and the microscopic. The macroscopic crystal may be regarded as a continuum, with symmetry properties defined by a system of intersecting axes and planes. Admissible rotations about the symmetry axes, and reflections through the symmetry planes, produce geometrical configurations equivalent to the original. The microscopic crystal is composed of atoms assembled into repeating patterns, which have symmetry properties defined by microscopic symmetry elements. These comprise screw axes and glide planes as well as symmetry axes and symmetry planes. However, a useful start into the theory is to ignore the screw axes and glide planes. Accordingly, in Part I, we construct patterns of atoms utilising only macroscopically available symmetry elements. This provides a natural foundation for later developments, including the full microscopic theory and the extension to colour symmetry. Also, with a slight adaptation of the terminology, the results of Part I are immediately relevant to the classification of crystal forms utilised in mineralogy and in other macroscopic applications.

1.2 SYMMETRY OPERATIONS

A fundamental notion in crystallography is that of axial symmetry. Thus, if we depict three identical atoms forming an equilateral triangle (Fig. 1.1), and

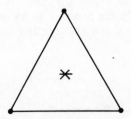

Fig. 1.1—The asterisk marks the intersection of the equilateral triangle with its 3-fold symmetry axis.

imagine an axis passing through its centre perpendicular to its plane, this will be a 3-fold symmetry axis for the triangle. Moie precisely, a rotation through $2\pi/3$ about the axis produces geometrical coincidence with the original configuration (Fig. 1.2). A second rotation through $2\pi/3$, i.e. a rotation

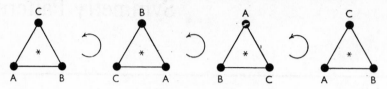

Fig. 1.2—Successive rotations of $\dfrac{2\pi}{3}$ about the principal symmetry axis of an equilateral triangle.

through $4\pi/3$ from the original configuration, again produces geometrical coincidence. A third iotation through $2\pi/3$, i.e. through 2π from the original configuration, restoies the original in every respect, being therefore equivalent to no rotation at all. Similar considerations hold for the square (4-fold symmetry axis) and for the regular hexagon (6-fold symmetry axis). More generally, the possession of an n-fold symmetry a ɔis means that a rotation through $2\pi/n$ about it produces geometrical coincidence. If so, there are in fact n distinct rotations about the axis which produce geometrical coincidence, i.e. through the angles

$$\frac{2\pi}{n}, \quad \frac{4\pi}{n}, \cdots \cdots, \quad \frac{2(n-1)\pi}{n}, \quad \frac{2n\pi}{n} \ (=2\pi), \tag{1}$$

the last being of course equivalent to no rotation at all.

It is not, perhaps, immediately obvious that the parallelogram has a symmetry axis, viz. a 2-fold axis passing through its centre perpendicular to its plane (Fig. 1.3). As regards the asymmetric quadrilateral, it could be said to have a 1-fold symmetry axis, because rotating it through $\dfrac{2\pi}{1}$ about any axis in space reproduces the original configuration. No other symmetry axes are admissible in mathematical crystallography. According to a simple but remarkable theorem (p. 70), the presence of an n-fold symmetry axis is not consistent with the existence of a space lattice (upon which all crystals are

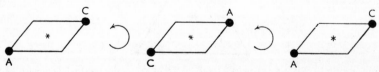

Fig. 1.3—Successive rotations of π about the principal axis of the parallelogram.

constructed) unless $n = 1$, 2, 3, 4, 6. No crystalline form has ever been discovered which contradicts this theorem, and it could therefore be used as a macroscopic argument in favour of the atomic theory of matter.

The possession of a symmetry axis perpendicular to its plane, usually termed a principal symmetry axis, by no means exhausts all the symmetry inherent in a regular polygon. For if we imagine planes passing through the principal axis, as indicated in Fig. 1.4, they will be symmetry planes for the

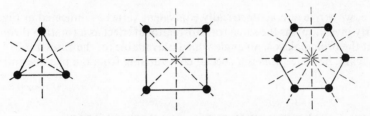

Fig. 1.4—Dashed lines mark the traces of symmetry planes perpendicular to the plane of the regular polygon (axial mirror planes).

polygon. More precisely, reflecting through any of these planes (axial symmetry planes) produces geometrical coincidence. Also, the plane coinciding with that of the polygon functions as a symmetry plane (transverse symmetry plane). However, this is a degenerate symmetry element since it transforms every atom into itself. Note that it intersects the axial planes in directions which are 2-fold symmetry axes for the polygon (Fig. 1.5), as can be seen by

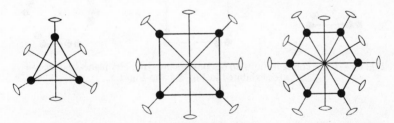

Fig. 1.5—The symbol ○——○ indicates a 2-fold secondary symmetry axis lying in the plane of the polygon.

rotating the plane of the polygon through π about these directions. These may be conveniently termed secondary axes by contrast with the principal axis. Clearly each secondary axis has the same transformational effect as the axial plane which intersects it, but this is a special effect which arises from the planar configuration.

The parallelogram, square, and regular hexagon are each symmetric with respect to an inversion through the centre. More precisely, if we join any vertex to the centre and continue this line by an equal length beyond the

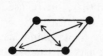

Fig. 1.6—The symbol $\longleftrightarrow$ connects atoms which are inverse with respect to the centre of each polygon.

centre, we arrive at a geometrically equivalent vertex as indicated in Fig. 1.6. Clearly inversion has the same transformational effect as a rotation through π about the principal axis, an angle which is available for the cases $n = 2, 4, 6$ of (1). Again, however, this is a special effect arising from the planar configuration.

1.3 ISOLATION OF PRINCIPAL SYMMETRY AXIS

A number of distinct symmetry elements have been identified for configurations forming regular polygons. To make further progress, it is necessary to isolate the principal symmetry axis and then combine it systematically with other available symmetry elements. Axial planes and secondary axes may be eliminated by introducing an asymetric motif unit in place of the single atom (Fig. 1.7). However, the transverse symmetry plane remains, and both the

Fig. 1.7—These configurations do not possess the symmetry planes and secondary axes exhibited in Fig. 1.4 and Fig. 1.5.

square and regular hexagon still possess inversion centres. Constructions are exhibited in Fig. 1.8 which eliminate all unwanted symmetries by virtue of the following features:

(a) A geometrical symbol specifying the relevant symmetry axis, imagined to pass through the symbol normal to the plane of the paper. This plane has the equation $z = 0$ in an obvious system of rectangular cartesian co-ordinates.

(b) A motif unit, conveniently depicted as a single atom, which is imagined to lie in the plane $z = 1$. This device allows us to introduce the plane $z = 0$ as a symmetry plane only if required. Also it allows us to introduce the point [0, 0, 0] as an inversion centre only if required.

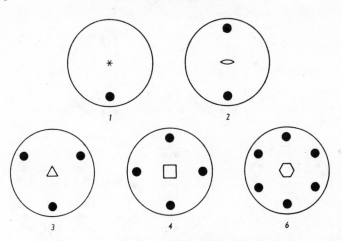

Fig. 1.8—Generation of a symmetry pattern from a motif unit (a dot representing a single atom) by the action of a principal symmetry axis.

(c) Replicas of the motif unit generated by the symmetry axis, the whole assembly of such units constituting a characteristic symmetry pattern. This is enclosed within a circle to emphasise its rotational significance.

(d) A mathematical symbol which serves to specify the symmetry element or, more actively, the symmetry pattern which it generates. Pure axial symmetries are symbolised *1, 2, 3, 4, 6* as the case may be.

1.4 COMBINATION OF SYMMETRY ELEMENTS

If a symmetry plane inteisects the symmetry axis, this combination provides new symmetries *1m, 2m, 3m, 4m, 6m* as displayed in Fig. 1.9. However, reference to Fig. 1.10 indicates the presence of further axial symmetry planes automatically introduced. Hence *2m, 4m, 6m* are conventionally written *2mm, 4mm, 6mm*. One *m* stands for the set of mirrors generated directly by the operation of the symmetry axis upon the original mirror; the second *m* stands for the interleaving mirrors inherently implied by the first set. On the other hand, *3m* stands unchanged since by virtue of the geometrical situation, the interleaving mirrors coincide with those generated directly. Finally, *1m* means nothing more than a single symmetry plane, which is conventionally written *m*.

If a symmetry plane lies transverse to the symmetry axis, this combination provides new symmetries $2/m, 3/m, 4/m, 6/m$ as displayed in Fig. 1.11, where the small circles indicate equivalent atoms in the plane $z = -1$. Of course $1/m$ means nothing more than a single symmetry plane, formally having a transverse orientation compared with *1m*. Such relationships are compactly expressed by writing $1/m \equiv 1m = m$.

If a principal symmetry axis is combined with a secondary symmetry axis, we obtain the new purely axial symmetries *22, 32, 42, 62* as displayed in

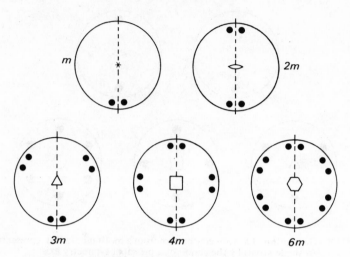

Fig. 1.9—Generation of symmetry patterns by a principal symmetry axis combined with an axial mirror plane.

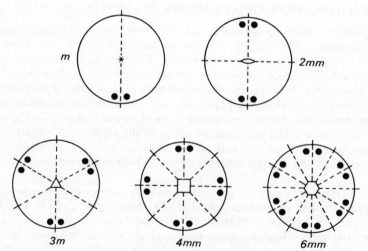

Fig. 1.10—These are the same configurations as in Fig. 1.9 but exhibiting the extra vertical mirror planes automatically introduced.

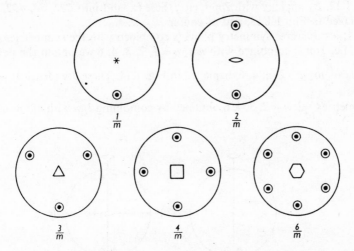

Fig. 1.11—It is necessary to think of the symbols ● and ○ as lying in the planes $z = 1$ and $z = -1$ respectively, being mirror images with respect to the plane $z = 0$.

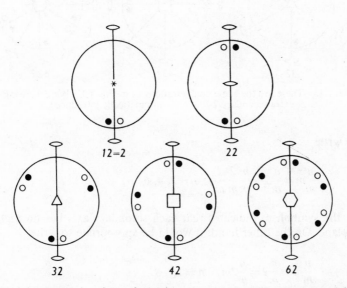

Fig. 1.12—Combination of a principal symmetry axis with a secondary 2-fold axis.

Fig. 1.12. By analogy with *2mm*, etc., these expand into *222*, *32*, *422*, *622* as displayed in Fig. 1.1.13, and of course *12 = 2*.

If a transverse symmetry plane is combined with the symmetries of Fig. 1.10, i.e. *1/m* is combined with *nm*; $n = 1, 2, 3, 4, 6$ we obtain the new symmetries $\frac{n}{m}m$; $n = 2, 3, 4, 6$ displayed in Fig. 1.14. These are identical with the symmetries $\frac{n}{m}2$; $n = 2, 3, 4, 6$ obtained by combining *1/m* with *n2*; $n = 2, 3, 4,$

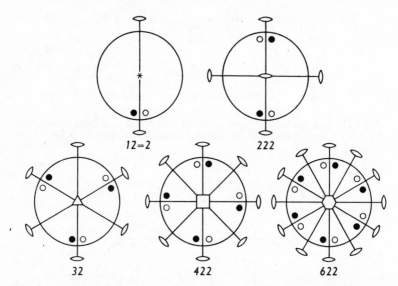

Fig. 1.13—These are the same configurations as in Fig. 1.12, but exhibiting the extra secondary 2-fold axes automatically introduced.

6. We write

$$\frac{n}{m}m = \frac{n}{m}2 = \frac{n}{m}\frac{2}{m}\frac{2}{m}; \quad n = 2, 4, 6 \tag{2}$$

where the symbolism indicates that each secondary axis lies normal to an axial plane. On the other hand, it would be appropriate to write

$$\frac{n}{m}m = \frac{n}{m}2 = \frac{n}{m}2m; \quad n = 1, \quad 3 \tag{3}$$

since in these cases each secondary axis is intersected by an axial plane. However, the case $n = 1$ does not yield a new symmetry since $\frac{1}{m}2m \equiv 2mm$.

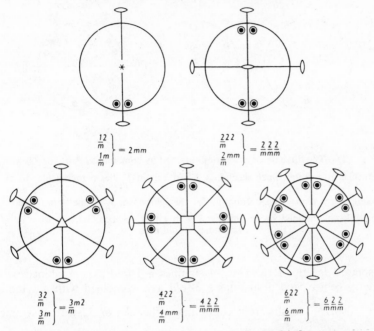

$$\left.\begin{array}{c}\frac{12}{m}\\\frac{1m}{m}\end{array}\right\}=2mm \qquad \left.\begin{array}{c}\frac{222}{m}\\\frac{2mm}{m}\end{array}\right\}=\frac{2\ 2\ 2}{m\ m\ m}$$

$$\left.\begin{array}{c}\frac{32}{m}\\\frac{3m}{m}\end{array}\right\}=\frac{3m2}{m} \qquad \left.\begin{array}{c}\frac{422}{m}\\\frac{4mm}{m}\end{array}\right\}=\frac{4\ 2\ 2}{m\ m\ m} \qquad \left.\begin{array}{c}\frac{622}{m}\\\frac{6mm}{m}\end{array}\right\}=\frac{6\ 2\ 2}{m\ m\ m}$$

Fig. 1.14—Combination of a transverse mirror plane with the configurations either of Fig. 1.13 $\left(\frac{1}{m}2, \frac{2}{m}22, \dots\right)$ or of Fig. 1.10 $\left(\frac{1}{m}m, \frac{2}{m}mm, \dots\right)$. Note that $\frac{1}{m}2 \left(=\frac{1}{m}m\right)$ is 2mm of Fig. 1.10 in a different orientation.

1.5 INVERSION CENTRE

There exists one point which remains fixed under all the symmetry operations so far listed, i.e. the point [0, 0, 0] which marks the common intersection of all the symmetry elements. This functions as an inversion centre for the symmetries

$$\frac{2}{m}, \frac{4}{m}, \frac{6}{m}; \frac{2}{m}\frac{2}{m}\frac{2}{m}, \frac{4}{m}\frac{2}{m}\frac{2}{m}, \frac{6}{m}\frac{2}{m}\frac{2}{m}, \qquad (4)$$

Since, as can be seen by inspection, any atom in these symmetry patterns is converted into an equivalent atom under the transformation

$$x, y, z \rightarrow -x, -y, -z \qquad (5)$$

The remaining symmetries do not possess inversion centres, a feature which suggests that we invert them with respect to [0, 0, 0] and so create new ones. However, only three new symmetries emerge from the exercise, viz.

$$\bar{1}, \ \bar{3}, \ \bar{3}\frac{2}{m} \qquad (6)$$

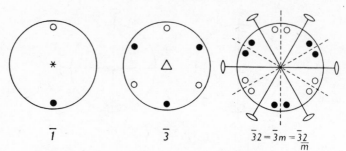

$$\overline{1} \qquad\qquad \overline{3} \qquad\qquad \overline{3}2=\overline{3}m=\overline{3}\frac{2}{m}$$

Fig. 1.15—The three new symmetries obtained by combining *1,3,3m* (or *32*) with an inversion centre. Each secondary 2-fold axis in $\overline{3}\frac{2}{m}$ lies perpendicular to an axial mirror plane, thus accounting for the symbolism $\overline{3}\frac{2}{m}$. Note that $\overline{3}\frac{2}{m}$ could have been obtained by combining $\overline{3}$ with either an axial mirror plane or a secondary 2-fold axis.

as displayed in Fig. 1.15. Thus, $\overline{1}$ signifies a 1-fold axis combined with a symmetry centre, and $\overline{3}$ signifies a 3-fold axis combined with a symmetry centre. Also, $\overline{3}\frac{2}{m}$ results from combining either *3m* or *32* with a symmetry centre, thereby automatically introducing secondary axes normal to axial planes as indicated by the symbolism. Clearly this symmetry may be alternatively constructed as $\overline{3}m$ or $\overline{3}2$, i.e.

$$\overline{3}m=\overline{3}2=\overline{3m}=\overline{32}=\overline{3}\frac{2}{m}\ .\tag{7}$$

By the same token, we have

$$\overline{1}m=\overline{1}2=(\overline{1m})=(\overline{12})=\overline{1}\frac{2}{m}\tag{8}$$

but of course $\quad \overline{1}\,\frac{2}{m}\equiv\frac{2}{m}\ .$

Using only descriptive geometrical methods, a total of 25 independent symmetry patterns have been constructed, which keep a point fixed under reflections, inversions, and admissible rotations. These constructions are summarised in Table 1.1. However, a mathematically preferable classification appears in Table 1.2.

Several questions may now be raised. Are the various symmetry elements (axis, plane, centre) all independent of each other? Do they suffice to describe all the symmetry properties of crystals? If not, how many more symmetries are there? These questions can be answered most simply by the formalism of group theory, which will be developed *ab initio* in the next chapter. Group theory is indispensable for a full understanding of the symmetries so far constructed as well as of all further extensions.

Table 1.1

axis	1	2	3	4	6
axis + transverse plane	$1/m(=m)$	$2/m$	$3/m$	$4/m$	$6/m$
axis + axial planes		$2mm$	$3m$	$4mm$	$6mm$
principal axis + secondary axes		222	32	422	622
axis + transverse plane + axial planes (or secondary axes)		$\dfrac{2\,2\,2}{mmm}$	$\dfrac{3}{m}2m$	$\dfrac{4\,2\,2}{mmm}$	$\dfrac{6\,2\,2}{mmm}$
previous symmetries + symmetry centre	$\bar{1}$		$\bar{3}$	$\bar{3}\dfrac{2}{m}$	

Table 1.2

pure axial symmetries (i.e. proper notations)	1	2 222	3 32 23	4 422 432	6 622
axial symmetries + symmetry centre (i.e. improper rotations involving symmetry centre)	$\bar{1}$	$2/m$ $\dfrac{2\,2\,2}{mmm}$	$\bar{3}$ $\bar{3}\dfrac{2}{m}$ $\dfrac{2}{m}\bar{3}^{*}$,	$4/m$ $\dfrac{4\,2\,2}{mmm}$ $\dfrac{4}{m}\bar{3}\dfrac{2}{m}^{*}$	$6/m$ $\dfrac{6\,2\,2}{mmm}$
symmetry planes not involving a symmetry centre (i.e. improper rotations not involving symmetry centre)	m	$2mm$ $\bar{4}^{*}$,	$3m$ $3/m$, $\dfrac{3}{m}2m$ $\bar{4}2m^{*}$,	$4mm$ $\bar{4}3m^{*}$	$6mm$

*From Table 2.1 and the symmetries (55), Chapter 2.

Chapter 2

Mathematical Formulation

2.1 CRYSTALLOGRAPHIC POINT GROUPS

Rotations, reflections and inversions may be described algebraically by crystallographic operators. These have matrix representations which enable us to determine their transformational effects (Appendix 1) but much of the theory can be developed without explicitly writing out any matrices. Suitable sets of operators constitute crystallographic point groups, i.e. finite groups corresponding with crystallographic symmetry patterns related to a fixed point. Examples of such patterns are provided in Chapter 1 and by the polyhedral forms often utilised to exhibit macroscopic crystalline symmetries [5, 6]. In this chapter and the next, we construct all the classical crystallographic point groups using elementary methods and results of finite group theory.

If the set of n crystallographic operators $\{G_1, G_2, \ldots, G_n\}$, for short written $\{G\}$, forms a group, then the following conditions must be satisfied: $\{G\}$ includes the identity operator I; the inverse of any operator belongs to the set, i.e. $G_i^{-1} \subset \{G\}$; products of operators belong to the set, i.e. $G_i G_j \subset \{G\}$; and the associative law holds for triple products, i.e.

$$G_i G_j . G_k = G_i G_j . G_k = G_i G_j . G_k. \tag{1}$$

These operators generate n equivalent atomic positions

$$G_i[x, y, z]; \qquad i = 1, 2, \ldots, n \tag{2}$$

from an arbitrary atomic position $[x, y, z]$, though of course degeneracy occurs if $[x, y, z]$ lies on a symmetry element. In particular if $[x, y, z]$ coincides with $[0, 0, 0]$, identified as the point of intersection of all the symmetry elements, this remains fixed under every operation. Clearly all the equivalent points (2) are equidistant from $[0, 0, 0]$, i.e.

$$|G_i[x, y, z]| = |[x, y, z]|; \qquad i = 1, 2, \ldots, n$$

which yields

$$\det G_i = \pm 1 \tag{3}$$

by standard matrix algebra procedures. The two cases $+1$, -1 of (3) correspond respectively with proper rotations, i.e. pure rotations such as can be directly undertaken by rigid bodies, and improper rotations, typified by reflections and inversions, which can only be simulated for rigid bodies through the presence of a mirror plane.

Instead of starting with a single atom, we may envisage a set of N atoms, of the same or different types, located at the N points

$$[x_r, y_r, z_r]; \qquad r = 1, 2, \ldots, N \qquad (4)$$

having no particular symmetry. This provides an asymmetric motif unit exemplified by the pair of atoms in Fig. 1.7. A replica of the configuration (4) is generated by G_i at the set of positions

$$G_i[x_r, y_r, z_r]; \qquad r = 1, 2, \ldots, N; \qquad (5)$$

and n equivalent replicas appear as we run through $i = 1, 2, \ldots n$, so producing a symmetry pattern which generalises that produced by a single atom. Nothing is gained mathematically by this generalisation, though of course it has considerable importance for the understanding of particular crystal structures.

2.2 CYCLIC GROUPS

If C stands for the matrix describing an axial rotation through $2\pi/n$, it follows that:

(a) C^2 describes an axial rotation through $4\pi/n$ and, more generally, C^r describes an axial rotation through $2\pi r/n$;

(b) C^n describes a rotation through 2π, and it must therefore be equivalent to the unit or identity matrix I. An obvious corollary is that $C^{n+r} = C^r$;

(c) C^{-1} describes a rotation through $2\pi/n$ in the opposite sense to that of C, and it must therefore be equivalent to C^{n-1}. More generally, $C^{-r} = C^{n-r}$.

All these properties of C imply that the set of operators

$$\{I, \quad C, \quad C^2, \ldots, C^{n-1}\}; \qquad C^n = I \qquad (6)$$

satisfies the conditions (1) above and therefore forms a group. This is the cyclic group of order n, conveniently symbolised $\{n\}$ in the present context, and we note that $\{1\}$, $\{2\}$, $\{3\}$, $\{4\}$, $\{6\}$ qualify as crystallographic point groups since they correspond with the symmetries $1, 2, 3, 4, 6$ of Chapter 1.

Mirror reflection can be described by an operator M with the properties

$$M^2 = I; \qquad M^{-1} = M. \qquad (7)$$

If so M generates the cyclic group

$$\{I, \quad M\}; \qquad M^2 = I, \tag{8}$$

symbolised $\{m\}$ since it corresponds with the crystallographic symmetry m. This group is mathematically isomorphic with the cyclic group $\{2\}$, though it describes a geometrically distinct symmetry, an equivalence conveniently expressed by writing $\{m\} \Leftrightarrow \{2\}$.

The inversion operator J is sufficiently defined by writing

$$J[x, y, z] = [-x, -y, -z] \tag{9}$$

where $[0, 0, 0]$ functions as the symmetry centre. If so, J generates the cyclic group

$$\{I, \quad J\}; \qquad J^2 = I \tag{10}$$

symbolised $\{\bar{I}\}$ since it corresponds with the crystallographic symmetry $\bar{I}$. This group is mathematically isomorphic with $\{m\}$ and $\{2\}$ though it describes a geometrically distinct symmetry from these, an equivalence usefully symbolised

$$\{\bar{I}\} \Leftrightarrow \{m\} \Leftrightarrow \{2\}. \tag{11}$$

2.3 ABELIAN GROUPS

We now couple a rotation C with a transverse mirror reflection M, so creating a rotoreflection operator MC (rotation followed by reflection) or CM (reflection followed by rotation). From the equality $CM = MC$, it follows that

$$MC^r = C^r M, \quad (MC^r)^{-1} = C^{-r} M^{-1} = MC^{n-r}; \quad C^n = I$$

$$(MC^r)^p = M^p C^{rp} = C^{rp}; \qquad p \text{ even}$$

$$\qquad\qquad = MC^{rp}; \qquad p \text{ odd} \tag{12}$$

where $MC, MC^2, \ldots, MC^{n-1}$ are distinct rotoreflection operators. These properties imply that the full set of $2n$ operators $\{n\} + M\{n\}$, i.e.

$$\left\{ \begin{array}{l} I, C, C^2, \ldots\ldots, C^{n-1} \\ M, MC, MC^2, \ldots\ldots, MC^{n-1} \end{array} \right\}; \quad C^n = M^2 = I, \tag{13}$$

forms an Abelian group since all the operators commute with each other. It is convenient to write

$$\{n\} + M\{n\} = \left\{ \frac{n}{m} \right\}; \quad n = 1, 2, 3, 4, 6 \tag{14}$$

because these groups correspond with the crystallographic symmetries

$$\frac{1}{m}(=m),\frac{2}{m},\frac{3}{m},\frac{4}{m},\frac{6}{m} \text{ respectively.}$$

The rotoinversion operator JC is analogous to MC, and every group constructed from MC has an isomorphic partner constructed from JC. In particular

$$\{n\}+J\{n\}=\{n\}+M\{n\}=\left\{\frac{n}{m}\right\}; \qquad n=2, 4, 6 \tag{15}$$

where the symbol '$=$' indicates both mathematical and geometrical equivalence. However this complete equivalence does not hold for n odd, and instead of (15) we write

$$\{n\}+J\{n\}=\{\bar{n}\}; \qquad n=1, 3 \tag{16}$$

since these groups correspond with the symmetries $\bar{1}, \bar{3}$. Utilising the symbol '$\Leftrightarrow$' already introduced in (11), we combine (16) and the cases $n=1, 3$ of (14) into the convenient display

$$\{\bar{n}\}=\{n\}+J\{n\}\Leftrightarrow\{n\}+M\{n\}=\left\{\frac{n}{m}\right\}; \qquad n=1, 3 \tag{17}$$

which supplements (15).

Cyclic groups may be generated from powers of MC, all of which—with one exception—reproduce those included in (17). Thus, if $C^n=I$ (n odd), we obtain the cyclic group of order $2n$:

$$\{MC\}_n=\left\{\begin{matrix} I, & MC, & M^2C^2, & \ldots\ldots, & M^{n-1}C^{n-1} \\ M^nC^n, & M^{n+1}C^{n+1}, & \ldots\ldots, & M^{2n-1}C^{2n-1} \end{matrix}\right\}; \quad M^2=C^n=I$$

i.e.

$$\{MC\}_n=\left\{\begin{matrix} I, & MC, & C^2, & MC^3, \ldots\ldots, & C^{n-1} \\ M, & C, & MC^2, & C^3, & \ldots\ldots, & MC^{n-1} \end{matrix}\right\}; \quad n \text{ odd} \tag{18}$$

which is seen to be identical with (13). Accordingly, the right-hand side of (17) expands into

$$\{n\}+M\{n\}=\left\{\frac{n}{m}\right\}=\{MC\}_n\Leftrightarrow\{2n\}; \qquad n=1, 3 \tag{19}$$

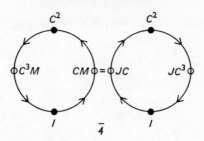

Fig. 2.1—The non-inversion symmetry $\bar{4}$ can be generated cyclically from powers of MC or, in opposite order, from powers of JC.

or more explicitly

$$\left\{\frac{1}{m}\right\}=\{MC\}_1\Leftrightarrow\{2\}, \qquad \left\{\frac{3}{m}\right\}=\{MC\}_3\Leftrightarrow\{6\}. \qquad (20)$$

If $C^n=I$ (n even), we obtain the cyclic group of order n:

$$\{MC\}_n=\{I, MC, M^2C^2, \ldots, M^{n-1}C^{n-1}\}; \qquad M^2=C^n=I,$$
$$=\{I, MC, C^2, \ldots, MC^{n-1}\}; \qquad n \text{ even}, \qquad (21)$$

which reproduces $\{\bar{1}\}$, $\{\bar{3}\}$ for $n=2$, 6. Accordingly, the left-hand side (17) expands into

$$\{\bar{n}\}=\{n\}+J\{n\}=\{MC\}_{2n}\Leftrightarrow\{2n\}; \qquad n=1, 3 \qquad (22)$$

or more explicitly

$$\{\bar{1}\}=\{MC\}_2\Leftrightarrow\{2\}, \qquad \{\bar{3}\}=\{MC\}_6\Leftrightarrow\{6\}. \qquad (23)$$

A special case arises when $n=4$, since (21) then yields

$$\{MC\}_4=\{I, \quad MC, \quad C^2, \quad MC^3\}; \qquad C^4=I \qquad (24)$$

corresponding with the non-inversion symmetry $\bar{4}$ exhibited in Fig. 2.1. Accordingly, we supplement (23) by writing

$$\{\bar{4}\}=\{MC\}_4\Leftrightarrow\{4\} . \qquad (25)$$

Analogous cyclic groups may be generated from powers of JC. Thus, replacing M by J in (18) reproduces $\{\bar{1}\}$, $\{\bar{3}\}$ for $n=1$, 3 as summarised by

$$\{\bar{1}\}=\{JC\}_1\Leftrightarrow\{2\} , \qquad \{\bar{3}\}=\{JC\}_3\Leftrightarrow\{6\} \qquad (26)$$

which parallels (20). Also, replacing M by J in (21) reproduces $\{1/m\}$, $\{\bar{4}\}$, $\{3/m\}$ as summarised by

$$\{1/m\} = \{JC\}_2 \Leftrightarrow \{2\}, \{\bar{4}\} = \{JC\}_4 \Leftrightarrow \{4\}, \{3/m\} = \{JC\}_6 \Leftrightarrow \{6\} \qquad (27)$$

which parallels (23), (25). This provides a more rational demarcation than previously since $\bar{4}$ now appears with the other non-inversion symmetries $1/m$, $3/m$. For this reason, possibly, rotoinversion is preferred to rotoreflection in modern treatments [5]. A useful symbolic device is to write $1/m = \bar{2}$, $3/m = \bar{6}$ so bringing these symmetries into line with $\bar{4}$. If so, (27) may be compactly displayed as

$$\{\bar{n}\} = \{JC\}_n \Leftrightarrow \{n\}; \qquad n = 2, 4, 6. \qquad (28)$$

The main results of this section may be summarised as follows:

$$\left\{\frac{n}{m}\right\} = \{n\} + M\{n\} = \{n\} + J\{n\}; \qquad n = 2, 4, 6$$

$$\{\overline{2n}\}, \text{ i.e. } \left\{\frac{n}{m}\right\} = \{n\} + M\{n\} = \{MC\}_n = \{JC\}_{2n}; \qquad n = 1, 3$$

$$\{\bar{n}\} = \{n\} + J\{n\} = \{MC\}_{2n} = \{JC\}_n; \qquad n = 1, 3$$

$$\{\bar{4}\} \qquad\qquad = \{MC\}_4 = \{JC\}_4 \qquad\qquad (29)$$

2.4 DIHEDRAL GROUPS

We now introduce a reflection operator D which satisfies (7), but refers to a mirror plane intersecting the symmetry axis (axial mirror). Coupling D with C provides n-1 composite operators

$$DC, DC^2, \ldots, DC^{n-1} \qquad\qquad (30)$$

which describe reflections through the other axial mirrors automatically created (Fig. 2.2). A feature of DC is that $DC \neq CD$, our first example of two non-commutative operators Reference to Fig. 2·2 shows that $CD = DC^{n-1}$, from which follow:

$$(DC)^2 = DC.DC = D.CD.C = D.DC^{n-1}.C = D^2C^n = I. \qquad (31)$$

These properties imply that the full set of $2n$ operators $\{n\} + D\{n\}$, i.e.

$$\left\{\begin{matrix} I, & C, & C^2, \ldots\ldots, & C^{n-1} \\ D, & DC, & DC^2, \ldots\ldots, & DC^{n-1} \end{matrix}\right\}; \quad C^n = D^2 = (CD)^2 = I \quad (32)$$

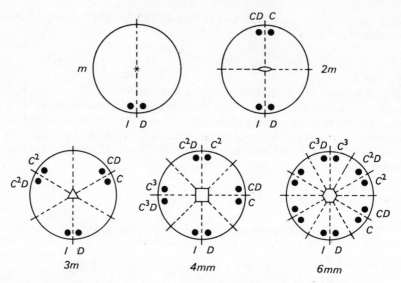

Fig. 2.2—Labelling of points by the composite operations which generate them from the one marked I, where D is an axial reflection operator.

forms a group termed the dihedral group. lt is convenient to write

$$\{n\} + D\{n\} = \{nm\}; \qquad n = 1, 2, 3, 4, 6 \tag{33}$$

since these groups correspond with the crystallographic symmetries *m, 2mm, 3m, 4mm, 6mm* respectively. Of course $\{1m\} \equiv \{1/m\} = \{m\}$. Also $\{2mm\} \Leftrightarrow \{2/m\}$ since $CD = DC$ in (32) for the case $n = 2$, which therefore becomes identical with (13) for the case $n = 2$.

All the mathematical properties of D remain unchanged, if we interpret it as describing a rotation through π about an axis intersecting the original axis at right angles. It is convenient to think of the original axis as the principal axis and the intersecting 2-fold axis as a secondary axis. If so, the $n-1$ composite operators (30) describe rotations about the other secondary axes automatically created (Fig. 2.3). In this case

$$\{n\} + D\{n\} = \{n2\}; \qquad n = 1, 2, 3, 4, 6 \tag{34}$$

since these groups correspond with the crystallographic symmetries *2, 222, 32, 422, 622* respectively. Our conclusions may be summarised as follows:

$$
\begin{aligned}
&\{222\} \Leftrightarrow \{2mm\} \Leftrightarrow \{2/m\} \qquad &\{12\} \Leftrightarrow \{1m\}, \text{ i.e. } \{2\} \Leftrightarrow \{m\} \\
&\{422\} \Leftrightarrow \{4mm\} \qquad &\{32\} \Leftrightarrow \{3m\} \\
&\{622\} \Leftrightarrow \{6mm\}
\end{aligned}
\tag{35}
$$

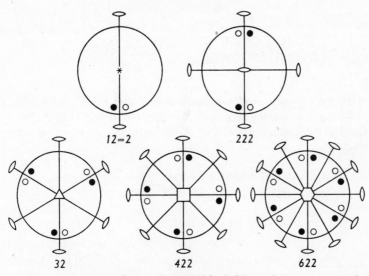

Fig. 2.3—Labels of points are similar to those in Fig. 2.2, but the operator D now signifies a rotation about secondary 2-fold axes.

The dihedral groups $\{\bar{n}2\}$, $\{\bar{n}m\}$ are geometrically and mathematically equivalent, i.e.

$$\{\bar{n}2\} = \{\bar{n}m\} = \{\bar{n}\} + D\{\bar{n}\}; \qquad n = 1, 2, 3, 4, 6 \qquad (36)$$

where D signifies either an axial symmetry plane or a secondary symmetry axis. These groups are displayed as follows in order to bring out their relationship with each other and their correspondence with the appropriate crystallographic symmetries:

$$\{\bar{2}m\} = \{\bar{2}2\} = \left\{\frac{1}{m}2m\right\} \equiv \{2mm\} \Leftrightarrow \{\bar{1}m\} = \{\bar{1}2\} = \left\{\bar{1}\frac{2}{m}\right\} \equiv \left\{\frac{2}{m}\right\}$$

$$\{\bar{6}m\} = \{\bar{6}2\} = \left\{\frac{3}{m}2m\right\} \qquad \Leftrightarrow \{\bar{3}m\} = \{\bar{3}2\} = \left\{\bar{3}\frac{2}{m}\right\} \qquad (37)$$

$$\{\bar{4}m\} = \{\bar{4}2\} = \{\bar{4}2m\}. \qquad (38)$$

Of these, only $\{\bar{4}2m\}$ has not previously been encountered in Chapter 1. It appears explicitly as

$$\{\bar{4}\} + D\{\bar{4}\},$$

i.e.

$$\left\{\begin{matrix} I, & MC, & C^2, & MC^3 \\ D, & DMC, & DC^2, & DMC^3 \end{matrix}\right\}; \quad C^4 = D^2 = (CD)^2 = I,$$

$$M^2 = I \qquad (39)$$

or

$$\left\{\begin{matrix} I, & JC, & C^2, & JC^3 \\ D, & DJC, & DC^2, & DJC^3 \end{matrix}\right\}; \qquad C^4 = D^2 = (CD)^2 = I,$$
$$J^2 = I \qquad\qquad (40)$$

where D, DC^2 signify axial symmetry planes and DMC, DMC^3 (or DJC, DJC^3) signify interleaving secondary axes, or the reverse (Fig. 1.14).

Only three further non-cubic crystallographic point groups may be constructed:

$$\begin{aligned} \{n2\} + J\ \{n2\} &= \{nm\} + J\ \{nm\} \\ \{n2\} + M\{n2\} &= \{nm\} + M\{nm\} \end{aligned} = \left\{\frac{n}{m}\frac{2}{m}\frac{2}{m}\right\}; \qquad n = 2, 4, 6. \quad (41)$$

However, for completeness we note that

$$\{n2\} + J\ \{n2\} = \{nm\} + J\ \{nm\} = \left\{\bar{n}\frac{2}{m}\right\}; \qquad n = 1, 3 \qquad (42)$$

$$\{n2\} + M\{n2\} = \{nm\} + M\{nm\} = \left\{\frac{n}{m}2m\right\}; \qquad n = 1, 3 \qquad (43)$$

so providing alternative constructions to (36) for these symmetries, which exactly correspond with the appropriate geometrical constructions (Figs. 1.14, 1.15) of Chapter 1. Finally, we note that nothing new emerges on replacing n by $\bar{n}$ in (41)–(43) since

$$\{\bar{n}2\} + J\{\bar{n}2\} = \{n2\} + J\{n2\}; \qquad n = 2, 4, 6$$
$$\{\bar{n}2\} + J\{\bar{n}2\} = \{n2\} + J\{n2\}; \qquad n = 1, 3 \qquad (44)$$

etc.

2.5 SUBGROUP STRUCTURE

Decompositions of the crystallographic point groups appear systematically in Chapter 4, where they play a fundamental role, but some key results may be usefully presented at this stage. If n is even, the set (32) can be re-demarcated as

$$\left\{\frac{n}{2}m\right\} + C\left\{\frac{n}{2}m\right\}; \qquad n = 2, 4, 6 \qquad (45)$$

where

$$\left\{\frac{n}{2}m\right\} = \{I, C^2, \ldots, C^{n-2}, D, DC^2, \ldots, DC^{n-2}\}; \qquad C^n = I$$

on bearing in mind $CD = DC^{n-1}$. This provides the new decompositions

$$\{nm\} = \left\{\frac{n}{2}m\right\} + C\left\{\frac{n}{2}m\right\}; \qquad n = 2, 4, 6 \tag{46}$$

$$\{n2\} = \left\{\frac{n}{2}2\right\} + C\left\{\frac{n}{2}2\right\}; \qquad n = 2, 4, 6 \tag{47}$$

associated with the alternative interpretations for D.

The sets (39), (40) can be re-demarcated on similar lines to provide the equivalent new decompositions

$$\{\bar{4}2m\} = \{2mm\} + MC\{2mm\} = \{222\} + MC\{222\}; \qquad C^4 = I \tag{48}$$

$$\{\bar{4}2m\} = \{2mm\} + JC\{2mm\} = \{222\} + JC\{222\}; \qquad C^4 = I \tag{49}$$

which supplement the decompositions (42), (43).

Starting with (13), it may be readily verified either geometrically or algebraically that

$$\left\{\frac{4}{m}\right\} = \left\{\frac{2}{m}\right\} + C\left\{\frac{2}{m}\right\} = \{\bar{4}\} + M\{\bar{4}\}; \qquad C^4 = I \tag{50}$$

$$\left\{\frac{2n}{m}\right\} = \left\{\frac{n}{m}\right\} + C\left\{\frac{n}{m}\right\} = \{\bar{n}\} + M\{\bar{n}\}; \qquad C^{2n} = I$$
$$n = 1, 3. \tag{51}$$

Also, immediate alternatives to the constructions (41) are

$$\left\{\frac{n}{m}\frac{2}{m}\frac{2}{m}\right\} = \left\{\frac{n}{m}\right\} + D\left\{\frac{n}{m}\right\}; \qquad D^2 = I$$
$$n = 2, 4, 6 \tag{52}$$

where D denotes either an axial mirror reflection or a secondary 2-fold rotation. Finally, it may be verified that

$$\left\{\frac{4}{m}\frac{2}{m}\frac{2}{m}\right\} = \left\{\frac{2}{m}\frac{2}{m}\frac{2}{m}\right\} + C\left\{\frac{2}{m}\frac{2}{m}\frac{2}{m}\right\} = \{\bar{4}2m\} + M\{\bar{4}2m\}; \; C^4 = I \tag{53}$$

$$\left\{\frac{6}{m}\frac{2}{m}\frac{2}{m}\right\} = \left\{\frac{3}{m}2m\right\} + C\left\{\frac{3}{m}2m\right\} = \left\{\bar{3}\frac{2}{m}\right\} + M\left\{\bar{3}\frac{2}{m}\right\}; \qquad C^6 = I. \tag{54}$$

2.6 CONCLUSIONS

Group theory has exposed two additional non-cubic symmetries not included in Table 1.1:

$\bar{4}$ $\bar{4}$-fold rotoinversion or rotoinflection axis

$\bar{4}2m$ $\bar{4}$ combined with either an axial symmetry plane or a secondary rotation axis   (55)

It has also exposed the four Abelian groups $\left\{\dfrac{n}{m}\right\}$, $\{\bar{n}\}$; $n = 1, 3$ as cyclic groups generated either by rotoinversion or rotoreflection operators. Finally it has exposed isomorphic equivalences between geometrically distinct symmetries, which are summarised in Table 2.1.

Table 2.1

---

cyclic	$1, 3$
	$2 \Leftrightarrow m \Leftrightarrow \bar{1}, 4 \Leftrightarrow \bar{4}, 6 \Leftrightarrow \dfrac{3}{m} \Leftrightarrow \bar{3}$
dihedral	$222 \Leftrightarrow 2mm \Leftrightarrow \dfrac{2}{m}, 422 \Leftrightarrow 4mm \Leftrightarrow \bar{4}2m$
	$622 \Leftrightarrow 6mm \Leftrightarrow \dfrac{3}{m}2m \Leftrightarrow \bar{3}\dfrac{2}{m}$
	$32 \Leftrightarrow 3m$
pure inversions	$4/m, \quad 6/m$
	$\dfrac{2}{m}\dfrac{2}{m}\dfrac{2}{m'}, \dfrac{4}{m}\dfrac{2}{m}\dfrac{2}{m'}, \dfrac{6}{m}\dfrac{2}{m}\dfrac{2}{m}$

This table comprises 27 distinct non-cubic crystallographic symmetries described by 14 mathematically independent point groups.

Chapter 3

Cubic Symmetries

3.1 INTRODUCTION

Crystallographic symmetry axes may be combined with each other in only six independent ways. Four of these have already been covered, i.e. *222*, *32*, *422*, *622* are each characterised by a principal axis and accompanying secondary axes orthogonal to it. The existence of these four may also be proved by an application of spherical trigonometry (Appendix 2), which yields two further possibilities, the tetrahedral symmetry *23* (short-hand for *222 3333*) and the octahedral symmetry *432* (short-hand for *444 3333 222222*). A convenient specialised notation for constructing these symmetries is as follows:

(a) $[hkl]$ denotes the symmetry axis having direction-rations $h:k:l$ relative to a system of rectangular cartesian co-ordinates;

(b) $(hkl)^{1/n}$ denotes the operator describing a rotation through $2\pi/n$ about $[hkl]$, in the sense of a right-hand screw motion (Fig. 3.1). Further operators generated by $(hkl)^{1/n}$ are

$$(hkl)^{1/n}, (hkl)^{2/n}, \ldots, (hkl)^{n/n} \quad \text{i.e. } I;$$

Fig. 3.1—Rotation operation $(hkl)^{1/n}$ about symmetry axis $[hkl]$.

(c) $\{pqr\}$ denotes the point-group corresponding with the symmetry pqr, a symbolism already utilised in Chapter 2. Also $G_i\{pqr\}$ denotes the set of operators obtained on left-multiplying each operator of $\{pqr\}$ by the operator G_i, and similarly for $\{pqr\}G_i$.

3.2 THE TETRAHEDRAL GROUP

We now combine a 3-fold axis [111] with a 2-fold axis [100], which are directions related as the diagonal and edge of a cube utilising an obvious system of rectangular cartesian co-ordinates. This combination implies two additional 2-fold axes [010], [001] and three additional 3-fold axes [$\overline{1}\overline{1}1$], [$1\overline{1}\overline{1}$], [$\overline{1}1\overline{1}$] as displayed in Fig. 3.2 all of which pass through the cube centre [0, 0, 0]. The relevant set of twelve operators.

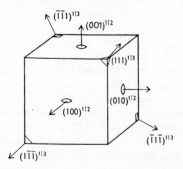

Fig. 3.2—Three 2-fold axes passing through the centre of a cube parallel to the cube edges define the symmetry *222*. These can be combined with the four 3-fold axes made up of the cube diagonals to yield the tetrahedral symmetry *23*.

$$\left\{\begin{array}{cccc} I & (100)^{1/2} & (010)^{1/2} & (001)^{1/2} \\ (111)^{1/3} & (\overline{1}\overline{1}1)^{1/3} & (1\overline{1}\overline{1})^{1/3} & (\overline{1}1\overline{1})^{1/3} \\ (111)^{2/3} & (\overline{1}\overline{1}1)^{2/3} & (1\overline{1}\overline{1})^{2/3} & (\overline{1}1\overline{1})^{2/3} \end{array}\right\} \qquad (1)$$

forms a group: it includes I; powers and inverses of operators belong to the set, e.g. $(111)^{-1/3} = (111)^{2/3}$; products of operators belong to the set, e.g.

$$(111)^{1/3} (100)^{1/2} = (\overline{1}\overline{1}1)^{1/3}, (100)^{1/2} (111)^{1/3} = (\overline{1}1\overline{1})^{1/3}$$

as may be verified from the matrix representations of Appendix 1; and the associative law holds for triple products. This is the tetrahedral group {23}, corresponding with the crystallographic symmetry *23* already mentioned. Notice that the first row of (1) itself forms a group, i.e. {*222*}. Also the second and third rows may be constructed as the products

$$\begin{aligned} (111)^{1/3}\{I, (100)^{1/2}, (010)^{1/2}, (001)^{1/2}\} &= (111)^{1/3}, (\overline{1}\overline{1}1)^{1/3}, (1\overline{1}\overline{1})^{1/3}, (\overline{1}1\overline{1})^{1/3} \\ (111)^{2/3}\{I, (100)^{1/2}, (010)^{1/2}, (001)^{1/2}\} &= (111)^{2/3}, (\overline{1}1\overline{1})^{2/3}, (\overline{1}\overline{1}1)^{2/3}, (1\overline{1}\overline{1})^{2/3} \end{aligned}$$

$$(2)$$

so enabling us to replace (1) by the more illuminating scheme

$$\{23\} = \{222\} + (111)^{1/3} \{222\} + (111)^{2/3} \{222\}. \tag{3}$$

A feature of the sets (2) is that

$$(111)^{1/3} \{222\} = \{222\} (111)^{1/3}, \; (111)^{2/3} \{222\} = \{222\}(111)^{2/3} \tag{4}$$

which may be exploited to provide an efficient proof that (3) forms a group (Appendix 3). This proof saves the 12×12 ($= 144$) products required to prove directly that (1) forms a group, though of course 4×4 ($= 16$) products are required to verify (4).

The symmetry *23* may be realised by placing four identical atoms at the alternate corners of a cube (Fig. 3.3). Since these atoms lie at the vertices of

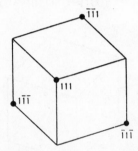

Fig. 3.3—Four atoms lying at alternative corners of a cube can be transformed into each other by operations of the tetrahedral group.

a regular tetrahedron, with centroid at the cube centre, it follows that $\{23\}$ comprises the twelve independent operators which rotate a regular tetrahedron into self-coincidence keeping its centroid fixed. These rotations are specified in detail in section 4 below. A more general arrangement exhibiting tetrahedral symmetry, accompanied by a stereogram (Phillips, Chap. II) is displayed in Fig. 3.4.

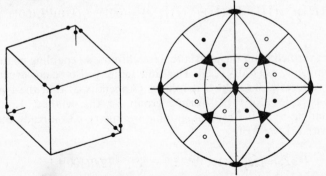

Fig. 3.4—The configuration of Fig. 3.3 is degenerate in that each atom lies on a 3-fold axis, a feature avoided here. No difficulty arises in correlating each atomic position with its stereographic projection shown on the right.

3.3 THE OCTAHEDRAL GROUP

Now we combine a 3-fold axis [111] with a 4-fold axis [100]. As before, this combination implies two additional 4-fold axes [010], [001] and three additional 3-fold axes $[\bar{1}11]$, $[1\bar{1}\bar{1}]$, $[\bar{1}1\bar{1}]$. However, there also appear six 2-fold axes

$$[011], \ [01\bar{1}], \ [101], \ [10\bar{1}], \ [110], \ [1\bar{1}0]$$

as displayed in Fig. 3.5. The relevant set of operators includes all the operators (1), plus the twelve additional operators

$$\left.\begin{array}{cccccc} (100)^{1/4} & (010)^{1/4} & (001)^{1/4} & (110)^{1/2} & (101)^{1/2} & (011)^{1/2} \\ (100)^{3/4} & (010)^{3/4} & (001)^{3/4} & (1\bar{1}0)^{1/2} & (10\bar{1})^{1/2} & (01\bar{1})^{1/2} \end{array}\right\} \quad (5)$$

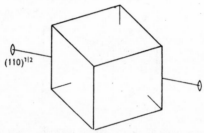

Fig. 3.5—Replacing the 2-fold axes of Fig. 3.2 by 4-fold axes automatically implies the introduction of six 2-fold axes $(110)^{1/2}, \ldots$ as shown.

These may be constructed as the twelve products

$$(100)^{1/4}\{I, (100)^{1/2}, (010)^{1/2}, (001)^{1/2}\} = \quad (100)^{1/4}, (100)^{3/4}, (011)^{1/2}, (01\bar{1})^{1/2}$$
$$(100)^{1/4}\{(111)^{1/3}, (\bar{1}\bar{1}1)^{1/3}, (1\bar{1}\bar{1})^{1/3}, (\bar{1}1\bar{1})^{1/3}\} = (101)^{1/2}, (010)^{3/4}, (10\bar{1})^{1/2}, (010)^{1/4}$$
$$(100)^{1/4}\{(111)^{2/3}, (\bar{1}\bar{1}1)^{2/3}, (1\bar{1}\bar{1})^{2/3}, (\bar{1}1\bar{1})^{2/3}\} = (001)^{3/4}, (1\bar{1}0)^{1/2}, (001)^{1/4}, (110)^{1/2}$$
$$(6)$$

as can be verified from the matrix representations, so enabling us to write (6) more compactly as $(100)^{1/4}$ {23}. The same set may also be constructed as the twelve products {23} $(100)^{1/4}$. It follows (Appendix 3) that the twenty-four operations {23}, $(100)^{1/4}$ {23} form a group, i.e. the octahedral group {432} corresponding with the crystallographic symmetry 432 already mentioned. In line with (3), we write

$$\{432\} = \{23\} + (100)^{1/4}\{23\} = \{23\} + \{23\}(100)^{1/4}. \quad (7)$$

The symmetry 432 may be realised by placing eight identical atoms at the eight corners of a cube, and it follows that {432} comprises the twenty-four independent operations which rotate a cube into self-coincidence keeping its centre fixed. These rotations are specified in detail below. A more general arrangement exhibiting octahedral symmetry, accompanied by a stereogram, is displayed in Fig. 3.6.

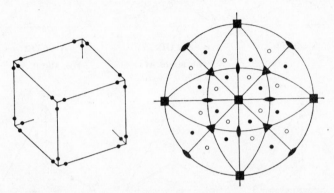

Fig. 3.6—Symmetry pattern exhibiting symmetry *432*. Partial degeneracy arises from the fact that atomic positions lie on the cube edges, a feature not assumed in the accompanying stereogram. Reference to Fig. 3.8(a) shows that this symmetry pattern also exhibits the symmetry $\dfrac{2}{m}\,\bar{3}$.

3.4 SYMMETRY OF REGULAR TETRAHEDRON AND CUBE

To determine the operations which transform a regular tetrahedron A, B, C, D into itself, its centroid G remaining fixed, we note that AG, BG, CG, DG constitute 3-fold symmetry axes for the tetrahedron. Of course AG passes through the centroid of the face BCD, etc. Accordingly, rotations of 0, $2\pi/3$, $4\pi/3$ (or $-2\pi/3$) about AG respectively generate the configurations

$$A : BCD, \quad A : DBC, \quad A : CDB$$

displayed in Fig. 3.7. The most useful next step is to rotate B into the position originally occupied by A, i.e. by a rotation through $2\pi/3$ about either CG or DG. Adopting the former, and keeping B fixed, we then generate the configurations

$$B : DCA, \quad B : ADC, \quad B : CAD$$

by rotations through 0, $2\pi/3$, $4\pi/3$ about BG, and similarly for the configurations

$$C : BDA, \quad C : ABD, \quad C : DAB$$
$$D : ACB, \quad D : BAC, \quad D : CBA.$$

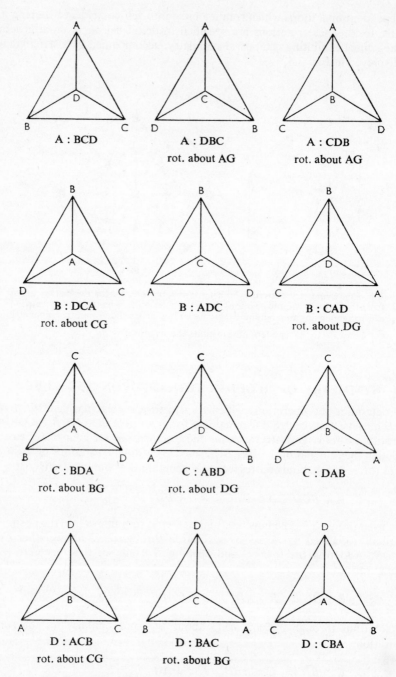

Fig. 3.7—The transformations of a regular tetrahedron into itself.

This procedure yields a total of $4 \times 3 = 12$ independent equivalent configurations, which fall into three distinct categories:

(a) the starting configuration $A : BCD$.

(b) the eight configurations

$$A : DBC, \quad A : CDB; \quad B : CAD, \quad B : DCA$$
$$C : BDA, \quad C : ABD; \quad D : ACB, \quad D : BAC$$

each of which has one vertex in common with $A : BCD$. For instance $B : CAD$, $C : ABD$ have D in common with $A : BCD$, and may therefore be generated from it by rotations of $\pm 2\pi/3$ about DG.

(c) the three configurations

$$B : ADC, \quad C : DAB, \quad D : CBA$$

each of which involves the simultaneous interchange of two neighbouring vertices relative to $A : BCD$. For instance $B : ADC$ was obtained above by the successive steps

$$A : BCD \rightarrow B : DCA \rightarrow B : ADC,$$

but it may also be generated directly by a rotation through π about an axis joining the mid-points of AB, CD. Since this axis passes through G, in a direction perpendicular to AB and CD, it constitutes a 2-fold symmetry axis for the tetrahedron, and similarly for the axes joining the mid-points of AC, BD (generating $C : DAB$) and AD, BC (generating $D : CBA$).

A regular tetrahedron therefore has four 3-fold symmetry axes and three 2-fold symmetry axes, all of which intersect at G, so providing twelve independent fixed-point rotations described by the operators (1).

The analysis of a cube having corners $A, B, C, D, A', B', C', D'$ proceeds on similar lines. Bearing in mind that the diagonals AA', etc., are 3-fold symmetry axes for the cube, we may generate three distinct equivalent configurations keeping AA' fixed. Of course AA' passes through the cube

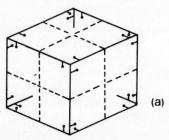

(a)

Fig. 3.8(a)—Triads related by reflection through the symmetry planes (001), (010) etc., yielding the symmetry $\frac{2}{m}\, \bar{3}$.

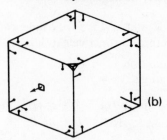

(b)—Triads related by rotations (100)¼, (010)¼ etc., yielding the symmetry *432*.

centre G. The corner B is then rotated into the position originally occupied by A, which can be done keeping G fixed, and we then generate three distinct equivalent configurations keeping BB' fixed. So proceeding for the remaining six corners, we generate a total of $8 \times 3 = 24$ independent configurations which fall into four distinct categories:

(a) the starting configuration.
(b) the eight configurations obtainable from (a) by rotations of $\pm 2\pi/3$ about the four diagonals.
(c) the nine configurations obtainable from (a) by rotations through $\pi/4$, $2\pi/4$, $3\pi/4$ about each of the three 4-fold axes (see Fig. 3.5).
(d) the six configurations obtainable from (a) by rotations through π about each of the six 2-fold axes (see Fig. 3.5).

All these symmetry axes intersect at G, so providing twenty-four independent fixed-point rotations described by the operators (1), (5).

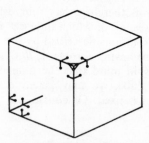

Fig. 3.9—Triads related by reflection through the symmetry planes (110) etc., yielding the symmetry $\bar{4}3m$.

3.5 INVERSION EFFECTS

The groups {23}, {432} may be expanded into the centro-symmetric crystallographic point-groups

$$\{23\} + J\{23\} = \left\{ \frac{2}{m} \bar{3} \right\}, \tag{8}$$

$$\{432\} + J\{432\} = \left\{\frac{4}{m} 3 \frac{2}{m}\right\}, \tag{9}$$

as follows from the fact that J commutes with every point-group operator (Appendix 3). The nature of (8), (9) may be readily understood by noting, firstly, that $\{23\}$, $\{432\}$ contain subgroups of the form

$$\{3\} = \{I, (hkl)^{1/3}, (hkl)^{2/3}\}; \quad hkl = 111, \text{ etc.}, \tag{10}$$

which expand into

$$\{3\} + J\{3\} = \{\bar{3}\}.$$

Also $\{23\}$ contains subgroups of the form

$$\{2\} = \{I, (hkl)^{1/2}\}; \quad hkl = 100, \text{ etc.}, \tag{11}$$

which expand into

$$\{2\} + J\{2\} = \left\{\frac{2}{m}\right\}.$$

Finally, $\{432\}$ contains subgroups of the form

$$\{2\} = \{I, (hkl)^{1/2}\}; \quad hkl = 110, 1\bar{1}0, \text{ etc.}, \tag{12}$$

$$\{4\} = \{I, (hkl)^{1/4}, (hkl)^{2/4}, (hkl)^{3/4}\}; \quad hkl = 100, \text{ etc.}, \tag{13}$$

which expand into

$$\{2\} + J\{2\} = \left\{\frac{2}{m}\right\}, \{4\} + J\{4\} = \left\{\frac{4}{m}\right\}.$$

These results demonstrate the existence of crystallographic symmetries $\frac{2}{m} \bar{3}$, $\frac{4}{m} 3 \frac{2}{m}$ built up from 23, 432 by the introduction of a symmetry centre. Geometrically expressed, the symmetry centre creates symmetry planes transverse to the 2-fold and 4-fold symmetry axes whilst also converting 3 into $\bar{3}$.

In parallel with (12), (13), there exist rotoinversion groups

$$\{\bar{2}\} \text{ i.e. } \{m\} = \{I, J(hkl)^{1/2}\}; \quad hkl = 110, 1\bar{1}0, \text{ etc.} \tag{14}$$

$$\{\bar{4}\} \qquad = \{I, J(hkl)^{1/4}, (hkl)^{2/4}, J(hkl)^{3/4}\}; \quad hkl = 100, \text{ etc.} \tag{15}$$

so allowing us to construct a new crystallographic point-group

$$\{\bar{4}3m\} = \{23\} + J(100)^{1/4}\{23\} \tag{16}$$

allied to $\{432\}$. In fact

$$\{\bar{4}3m\} \Leftrightarrow \{432\} \tag{17}$$

since $\{m\} \Leftrightarrow \{2\}$, $\{\bar{4}\} \Leftrightarrow \{4\}$.

This defines a new crystallographic symmetry $\bar{4}3m$ allied to 432, and our analysis shows that replacing 4 by $\bar{4}$ in 432 automatically replaces 2 by m.

Stereograms of the symmetries $\frac{2}{m}\bar{3}$, $\frac{4}{m}\bar{3}\frac{2}{m}$, $\bar{4}3m$ are displayed in Fig. 3.10. There exist no further cubic possibilities. Accordingly, Table 3.1 must be supplemented as follows.

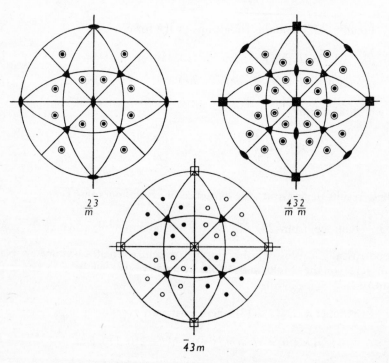

$\frac{2}{m}\bar{3}$

$\frac{4}{m}\bar{3}\frac{2}{m}$

$\bar{4}3m$

Fig. 3.10—Stereograms of the symmetries $\frac{2}{m}\bar{3}$, $\frac{4}{m}\bar{3}\frac{2}{m}$, $\bar{4}3m$.

Table 3.1

tetrahedral:	*23*
octahedral:	$432 \Leftrightarrow \bar{4}3m$
centro-symmetric:	$\dfrac{2}{m}\bar{3}, \ \dfrac{4}{m}\bar{3}\dfrac{2}{m}$

This Table comprises 5 distinct crystallographic symmetries described by four mathematically independent point groups. Combining Tables 2.1 and 3.1, we obtain thirty-two distinct crystallographic symmetries described by eighteen mathematically independent point groups For ease of reference, all these symmetries have already been incorporated into Table 1.2.

3.6 DEGENERACIES

It may be remarked that the symmetry $\dfrac{2}{m}\bar{3}$, as well as *432*, is obeyed by the configuration in Fig. 3.6. To discriminate between these, we introduce a more asymmetrical triad of atoms at the cube corner. According to $\dfrac{2}{m}\bar{3}$, the arrangements at neighbouring corners result from symmetry planes parallel to the cube faces as depicted in Fig. 3.8a. According to *432*, they result from 4-fold symmetry axes normal to the cube faces as depicted in Fig. 3.8b. Clearly these alternative configurations become identical for atoms on the cube edges, so yielding the degenerate configuration of Fig. 3.6. Similar remarks hold for Fig. 3.4, which obeys the symmetry $\bar{4}3m$ as well as *23*. Utilising the above asymmetrical triad, we obtain a non-degenerate configuration obeying $\bar{4}3m$ (Fig. 3.9). This, of course, also obeys *23* since $\{23\}$ is a subgroup of $\{\bar{4}3m\}$.

Chapter 4

Colour Point Groups

4.1 COLOUR SYMMETRY

The symmetry patterns so far considered have been entirely geometrical. However, we may construct patterns involving colour symmetry which provide a consistent extension of geometrical symmetry. Atoms do not of course possess colour but in some crystals they possess magnetic moments of positive or negative orientation corresponding mathematically with two distinct colours*. Let S denote the operator which switches colour, say, from black to white or vice versa. Then clearly

$$S^2 = I, \qquad S^{-1} = S. \tag{1}$$

Also S commutes with every point-group operator since colour transformations do not affect geometrical transformations, i.e.

$$SG_i = G_i S; \qquad G_i \subset \{G\} \tag{2}$$

where $\{G\}$ denotes any one of the thirty-two crystallographic point-groups. An immediate implication of (1), (2) is that (Appendix 3) we can always construct a colour group

$$\{G\} + S\{G\} \tag{3}$$

associated with $\{G\}$. These expanded groups are termed 'grey' groups since they colour every equivalent atom both black and white. However, such groups do not provide any essentially new symmetries.

A systematic procedure for creating genuinely new colour symmetries is to take every $\{G\}$ of even order $2n$, and if possible write

$$\text{where} \qquad \begin{matrix} \{G\} = \{H\} + k\{H\}; & k \subset \{G\}k \notin \{H\} \\ \{H\} = \text{a subgroup of } \{G\} \text{ of order } n \end{matrix} \Bigg\}. \tag{4}$$

---

* Attention should be drawn to the internal dynamic symmetries of a crystal which may also be described by colour space groups, as first proposed by B. Kolakowski, [23] and papers referenced there.

This decomposition enables us to construct the colour group $\{G'\}$ of order $2n$ isomorphic with $\{G\}$:

$$\left.\begin{array}{l} \{G'\}=\{H\}+Sk\{H\} \\ \{G'\}\Leftrightarrow\{G\} \end{array}\right\}. \tag{5}$$

where

The decomposition is not necessarily unique, and for each distinct choice of $\{H\}$ in (4) there corresponds a distinct $\{G'\}$. According to a theorem of B. M. Hurley (Appendix 3), the maximum number of decompositions of the type (4) is given by:

$$2^r-1; \qquad r=\text{number of independent generators of } \{G\}. \tag{6}$$

However the actual number of decompositions can only be found from an inspection of each group following the categories already laid down. This programme will now be implemented.

4.2 CYCLIC COLOUR GROUPS

Since every cylic group can be generated from a single operator, formula (6) allows at most one decomposition, which always exists for cyclic groups of even order. Thus, associated with $\{2\}, \{4\}, \{6\}$ we construct the colour groups

$$\{2'\}=\{1\}+SC\{1\}; \quad C^2=I \tag{7}$$
$$\{4'\}=\{2\}+SC\{2\}; \quad C^4=I \tag{8}$$
$$\{6'\}=\{3\}+SC\{3\}; \quad C^6=I \tag{9}$$

displayed in Fig. 4.1. As might be expected, these are cyclic groups generated from powers of SC, i.e.

$$\{n'\}=\{SC\}_n; \quad (SC)^n=S^nC^n=C^n=I \tag{10}$$
$$n=2,4,6$$

following the symbolism of (29), Chap. 2.

Associated with $\{\bar{n}\}, \{2\bar{n}\}; \; n=1, 3$ we construct the colour groups

$$\{\bar{1}'\}=\{1\}+SJ\{1\}, \quad \{\bar{3}'\}=\{3\}+SJ\{3\} \tag{11}$$

$$\{\bar{2}'\}=\left\{\frac{1}{m'}\right\}=\{1\}+SM\{1\}, \quad \{\bar{6}'\}=\left\{\frac{3}{m'}\right\}=\{3\}+SM\{3\} \tag{12}$$

which are cylic groups generated from powers of SMC or SJC, i.e.

$$\{\bar{n}'\}=\{SMC\}_{2n}=\{SJC\}_n; \quad n=1, 3 \tag{11a}$$
$$\{2\bar{n}'\}=\{SMC\}_n=\{SJC\}_{2n}; \quad n=1, 3. \tag{12a}$$

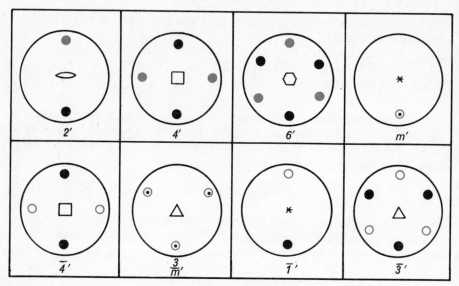

Fig. 4.1—Cyclic colour symmetries.

Finally, following (8), we obtain

$$\{\overline{4}'\} = \{2\} + SMC\{2\} = \{2\} + SJC\{2\}; \quad C^4 = I, \tag{13}$$
$$= \{SMC\}_4 = \{SJC\}_4.$$

All these symmetries are displayed in Fig. 4.1.

4.3 DIHEDRAL COLOUR GROUPS

Dihedral groups are generated from two operators so allowing a maximum of three decompositions according to formula (6). In fact there exist two decompositions for $\{nm\}$; $n = 2, 4, 6$ following (33), (46) Chap. 2. Thus

$$\{2mm\} = \{2\} + D\{2\} = \{I, C, D, DC\}; \quad C^2 = D^2 = 1$$
$$= \{m\} + C\{m\} = \{I, D, C, CD\}; \quad CD = DC$$

where C signifies a 2-fold rotation operator, D signifies an axial mirror reflection, and DC signifies a second axial mirror reflection. Accordingly we may construct the two independent colour groups conveniently symbolised

$$\{2m'm'\} = \{2\} + SD\{2\} = \{I, C, SD, SDC\} \tag{14}$$
$$\{2'mm'\} = \{m\} + SC\{m\} = \{I, D, SC, SCD\} \tag{15}$$

where SD, SDC signify colour reflections and SC signifies a colour rotation. Similarly we obtain

$$\{4m'm'\} = \{4\} + SD\{4\} \tag{16}$$
$$\{4'mm'\} = \{2mm\} + SC\{2mm\}; \quad C^4 = I \tag{17}$$

and

$$\{6m'm'\} = \{6\} + SD\{6\} \tag{18}$$
$$\{6'mm'\} = \{3m\} + SC\{3m\}; \quad C^6 = I. \tag{19}$$

However, only one decomposition exists for $\{nm\}$; $n = 1, 3$ yielding the colour, groups

$$\{1m'\} = \{1\} + SD\{1\} \equiv \left\{\frac{1}{m'}\right\} = \{m'\} \tag{20}$$
$$\{3m'\} = \{3\} + SD\{3\}. \tag{21}$$

These symmetries are displayed in Fig. 4.2.

The mathematical equivalence $\{n2\} \Leftrightarrow \{nm\}$ has already been noted, which implies that new colour groups may be constructed by interpreting D as a secondary rotation through π in (14)–(21). Thus we now suppose $D = (100)^{1/2}$ whilst of course $C = (001)^{1/n}$; $n = 2, 3, 4, 6$ as before. If so the two independent colour groups (14), (15) become

$$\{22'2'\} = \{I, (001)^{1/2}, S(100)^{1/2}, S(010)^{1/2}\} \tag{22}$$
$$\{2'22'\} = \{I, (100)^{1/2}, S(001)^{1/2}, S(010)^{1/2}\} \tag{23}$$

which correspond with the geometrically equivalent colour symmetries $22'2'$, $2'22'$. No complications arise with the analogues of (16)–(21):

$$\{42'2'\} = \{4\} + SD\{4\} \tag{24}$$
$$\{4'22'\} = \{222\} + SC\{222\}; \quad C^4 = I \tag{25}$$
$$\{62'2'\} = \{6\} + SD\{6\} \tag{26}$$
$$\{6'22'\} = \{32\} + SC\{32\}; \quad C^6 = I \tag{27}$$
$$\{12'\} = \{1\} + SD\{1\} = \{2'\} \tag{28}$$
$$\{32'\} = \{3\} + SD\{3\}. \tag{29}$$

These symmetries are displayed in Fig. 4.2.

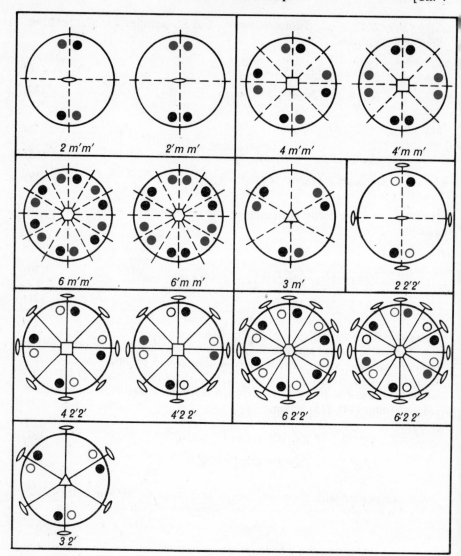

Fig. 4.2—Dihedral colour symmetries.

Three independent decompositions exist for each of the dihedral groups $\left\{\bar{3}\frac{2}{m}\right\}$, $\left\{\frac{3}{m}2m\right\}$ defined by (37), Chap. 2, i.e. the initial decomposition plus the constructions (42), (43). These yield the colour groups

$$\left\{\bar{3}\frac{2'}{m'}\right\} = \{\bar{3}\} + SD\{\bar{3}\} \tag{30}$$

$$\left\{\bar{3}'\frac{2}{m'}\right\} = \{32\} + SJ\{32\} \tag{31}$$

$$\left\{\bar{3}'\frac{2'}{m}\right\} = \{3m\} + SJ\{3m\} \tag{32}$$

$$\left\{\frac{3}{m}2'm'\right\} = \left\{\frac{3}{m}\right\} + SD\left\{\frac{3}{m}\right\} \tag{33}$$

$$\left\{\frac{3}{m'}2m'\right\} = \{32\} + SM\{32\} \tag{34}$$

$$\left\{\frac{3}{m'}2'm\right\} = \{3m\} + SM\{3m\} \tag{35}$$

where D signifies either an axial mirror reflection or a secondary rotation through π, and M denotes a transverse mirror reflection. Similarly there exists three decompositions for $\{\bar{4}2m\}$, defined in (38), Chap. 2, i.e. the initial construction plus the decompositions (48) or (49). These yield the colour groups

$$\{\bar{4}2'm'\} = \{\bar{4}\} + SD\{\bar{4}\} \tag{36}$$

$$\{\bar{4}'2m'\} = \{222\} + SMC\{222\} \quad C^4 = I \tag{37}$$
$$\{222\} + SJC\{222\};$$

$$\{\bar{4}'2'm\} = \{2mm\} + SMC\{2mm\} \quad C^4 = I. \tag{38}$$
$$\{2mm\} + SJC\{2mm\};$$

Of the remaining dihedral groups in (37), Chap. 2, it may be noted that $\left\{\bar{1}\frac{2}{m}\right\}$ is formally analogous to $\left\{\bar{3}\frac{2}{m}\right\}$ and so yields the colour groups

$$\left\{\bar{1}\frac{2'}{m'}\right\} = \{\bar{1}\} + SD\{\bar{1}\} \equiv \left\{\frac{2'}{m'}\right\} \tag{39}$$

$$\left\{\bar{1}'\frac{2}{m'}\right\} = \{12\} + SJ\{12\} \equiv \left\{\frac{2}{m'}\right\} \tag{40}$$

$$\left\{\bar{1}'\frac{2'}{m}\right\} = \{1m\} + SJ\{1m\} \equiv \left\{\frac{2'}{m}\right\} \tag{41}$$

analogous to (30)–(32). However $\left\{\bar{1}\frac{2}{m}\right\}$ is equivalent to $\left\{\frac{2}{m}\right\}$ in a different orientation, so allowing an alternative treatment in the next section. The final group $\left\{\frac{1}{m}2m\right\}$ of (37), Chap. 2, is formally analogous to $\left\{\frac{3}{m}2m\right\}$ and so yields the colour groups

$$\left\{\frac{1}{m}2'm'\right\} = \left\{\frac{1}{m}\right\} + SD\left\{\frac{1}{m}\right\} \equiv \{2'mm'\} \tag{42}$$

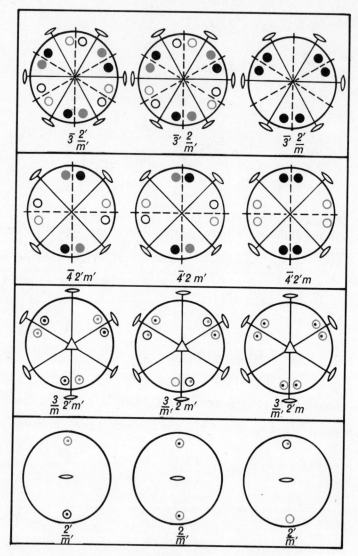

Fig. 4.3—Further dihedral colour symmetries.

$$\left\{\frac{1}{m'}2m'\right\} = \{12\} + SM\{12\} \equiv \{2m'm'\} \tag{43}$$

$$\left\{\frac{1}{m'}2'm\right\} = \{1m\} + SM\{1m\} \equiv \{2'm'm\} \tag{44}$$

analogous to (33)–(35). Clearly (44) is geometrically equivalent to (42), so providing only two independent colour groups in this case—in line with the results (14), (15) since $\left\{\dfrac{1}{m}2m\right\} \equiv \{2mm\}$. The various symmetries are displayed in Fig. 4.3.

4.4 ABELIAN COLOUR GROUPS

The Abelian groups $\left\{\dfrac{n}{m}\right\}$; $n = 2, 4, 6$ are each generated by two operators, so allowing a maximum of three decompositions which always exist, i.e. the initial constructions (15), Chap. 2, plus the decompositions (50), (51) Chap. 2. These yield the colour groups

$$\left\{\frac{2}{m'}\right\} = \{2\} + SM\{2\} = \{2\} + SJ\{2\} \tag{45}$$

$$\left\{\frac{2'}{m}\right\} = \left\{\frac{1}{m}\right\} + SC\left\{\frac{1}{m}\right\}; \quad C^2 = I \tag{46}$$

$$\left\{\frac{2'}{m'}\right\} = \{\bar{1}\} + SM\{\bar{1}\} \tag{47}$$

$$\left\{\frac{6}{m'}\right\} = \{6\} + SM\{6\} = \{6\} + SJ\{6\} \tag{48}$$

$$\left\{\frac{6'}{m}\right\} = \left\{\frac{3}{m}\right\} + SC\left\{\frac{3}{m}\right\}; \quad C^6 = I \tag{49}$$

$$\left\{\frac{6'}{m'}\right\} = \{\bar{3}\} + SM\{\bar{3}\} \tag{50}$$

$$\left\{\frac{4}{m'}\right\} = \{4\} + SM\{4\} = \{4\} + SJ\{4\} \tag{51}$$

$$\left\{\frac{4'}{m}\right\} = \left\{\frac{2}{m}\right\} + SC\left\{\frac{2}{m}\right\}; \quad C^4 = I \tag{52}$$

$$\left\{\frac{4'}{m'}\right\} = \{\bar{4}\} + SM\{\bar{4}\}. \tag{53}$$

These symmetries are displayed in Fig. 4.4, and of course (45)–(47) have already been constructed in (39)–(41) from the decomposition of $\left\{\bar{1}\dfrac{2}{m}\right\}$.

4.5 EXTENDED DIHEDRAL COLOUR GROUPS

The groups $\left\{\dfrac{n}{m}\dfrac{2}{m}\dfrac{2}{m}\right\}$; $n = 2, 4, 6$ are each generated from three operators so allowing a maximum of seven decompositions according to (6). However, there exist only three decompositions for $\left\{\dfrac{2}{m}\dfrac{2}{m}\dfrac{2}{m}\right\}$, i.e. the initial construc-

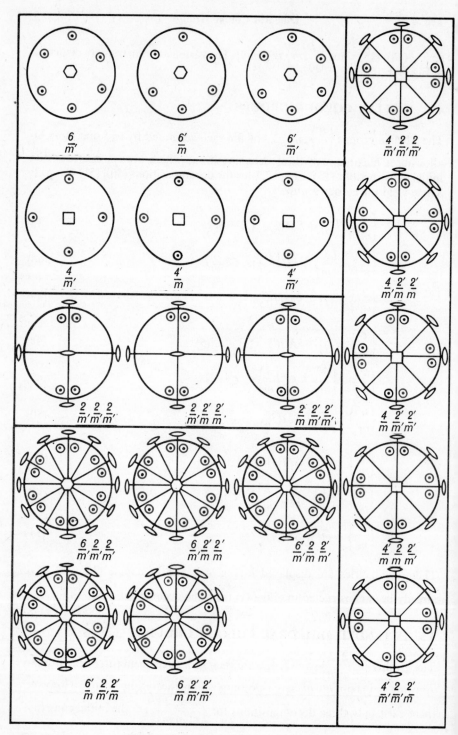

Fig. 4.4—Colour symmetries with inversion.

tions (41), Chap. 2, plus the decomposition (52), Chap. 2, yielding the colour groups

$$\left\{\frac{2}{m'}\frac{2}{m'}\frac{2}{m'}\right\}=\{222\}+SM\{222\}=\{222\}+SJ\{222\} \tag{54}$$

$$\left\{\frac{2}{m'}\frac{2'}{m}\frac{2'}{m}\right\}=\{2mm\}+SM\{2mm\}=\{2mm\}+SJ\{2mm\} \tag{55}$$

$$\left\{\frac{2}{m}\frac{2'}{m'}\frac{2'}{m'}\right\}=\left\{\frac{2}{m}\right\}+SD\left\{\frac{2}{m}\right\}. \tag{56}$$

Five decompositions exist for $\left\{\frac{4}{m}\frac{2}{m}\frac{2}{m}\right\}$, i.e. the initial constructions (41), Chap. 2, plus the decompositions (52), (53) of Chap. 2, yielding the colour groups

$$\left\{\frac{4}{m'}\frac{2}{m'}\frac{2}{m'}\right\}=\{422\}+SM\{422\}=\{422\}+SJ\{422\} \tag{57}$$

$$\left\{\frac{4}{m'}\frac{2'}{m}\frac{2'}{m}\right\}=\{4mm\}+SM\{4mm\}=\{4mm\}+SJ\{4mm\} \tag{58}$$

$$\left\{\frac{4}{m}\frac{2'}{m'}\frac{2'}{m'}\right\}=\left\{\frac{4}{m}\right\}+SD\left\{\frac{4}{m}\right\} \tag{59}$$

$$\left\{\frac{4'}{m}\frac{2}{m}\frac{2'}{m'}\right\}=\left\{\frac{2}{m}\frac{2}{m}\frac{2}{m}\right\}+SC\left\{\frac{2}{m}\frac{2}{m}\frac{2}{m}\right\}; \quad C^4=I \tag{60}$$

$$\left\{\frac{4'}{m'}\frac{2}{m'}\frac{2'}{m}\right\}=\{\bar{4}2m\}+SM\{\bar{4}2m\}. \tag{61}$$

Similarly there exist five decompositions for $\left\{\frac{6}{m}\frac{2}{m}\frac{2}{m}\right\}$, yielding the colour groups

$$\left\{\frac{6}{m'}\frac{2}{m'}\frac{2}{m'}\right\}=\{622\}+SM\{622\}=\{622\}+SJ\{622\} \tag{62}$$

$$\left\{\frac{6}{m'}\frac{2'}{m}\frac{2'}{m}\right\}=\{6mm\}+SM\{6mm\}=\{6mm\}+SJ\{6mm\} \tag{63}$$

$$\left\{\frac{6}{m}\frac{2'}{m'}\frac{2'}{m'}\right\}=\left\{\frac{6}{m}\right\}+SD\left\{\frac{6}{m}\right\} \tag{64}$$

$$\left\{\frac{6'}{m}\frac{2}{m'}\frac{2'}{m}\right\}=\left\{\frac{3}{m}2m\right\}+SC\left\{\frac{3}{m}2m\right\}; \quad C^6=I \tag{65}$$

$$\left\{\frac{6'}{m'}\frac{2}{m}\frac{2'}{m'}\right\}=\left\{\bar{3}\frac{2}{m}\right\}+SM\left\{\bar{3}\frac{2}{m}\right\}. \tag{66}$$

Note that the distribution of primed symbols follows the same pattern in (54)–(56), (57)–(59) and (62)–(64). However, it differs slightly from (60), (61) to (65), (66). The various symmetries are displayed in Fig. 4.4.

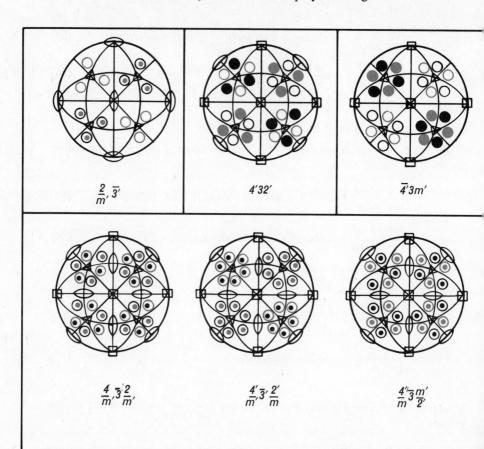

Fig. 4.5—Stereograms of cubic colour symmetries.

4.6 CUBIC COLOUR GROUPS

By virtue of (4), Chap. 3, the tetrahedral group $\{23\}$ does not have a decomposition of the type (4). However this group may be used to build the colour groups

$$\left\{ \frac{2}{m'} \,\overline{3}' \right\} = \{23\} + SJ\{23\} \tag{67}$$

$$\{4'32'\} = \{23\} + S(100)^{1/4}\{23\} \tag{68}$$

$$\{\overline{4}'3m'\} = \{23\} + SJ(100)^{1/4}\{23\}. \tag{69}$$

Finally, $\left\{\dfrac{4}{m}\, \bar{3}\, \dfrac{2}{m}\right\}$ yields the colour groups

$$\left\{\frac{4}{m'}\, \bar{3}'\, \frac{2}{m'}\right\}=\{432\}+SJ\{432\} \tag{70}$$

$$\left\{\frac{4'}{m'}\, \bar{3}'\, \frac{2'}{m}\right\}=\{\bar{4}3m\}+SJ\{\bar{4}3m\} \tag{71}$$

$$\left\{\frac{4'}{m}\, \bar{3}\, \frac{2'}{m'}\right\}=\left\{\frac{2}{m}\, \bar{3}\right\}+S(100)^{1/4}\left\{\frac{2}{m}\, \bar{3}\right\}. \tag{72}$$

Stereograms of the cubic colour symmetries are displayed in Fig. 4.5.

4.7 CONCLUSIONS

Building upon the thirty-two classical crystallographic point groups, a total of fifty-eight independent colour crystallographic point groups have been constructed. These are listed in Table 4.1.

Classical	Colour
2	2'
4	4'
6	6'
$\bar{1}$	$\bar{1}'$
$\bar{3}$	$\bar{3}'$
$\bar{2}\,(=m)$	$\bar{2}'\,(=m')$
$\bar{6}\left(=\dfrac{3}{m}\right)$	$\bar{6}'\left(=\dfrac{3}{m'}\right)$
$\bar{4}$	$\bar{4}'$
2mm	2m'm', 2'mm'
4mm	4m'm', 4'mm'
6mm	6m'm', 6'mm'
3m	3m'
222	22'2'
422	42'2', 4'22'
622	62'2', 6'22'
32	32'
$3\dfrac{2}{m}$	$\bar{3}\dfrac{2'}{m'}$, $\bar{3}'\dfrac{2}{m'}$, $\bar{3}'\dfrac{2'}{m}$

Classical	Colour
$\dfrac{3}{m}2m$	$\dfrac{3}{m}2'm'$, $\dfrac{3}{m'}2m'$, $\dfrac{3}{m'}2'm$
$\bar{4}2m$	$\bar{4}2'm'$, $\bar{4}'2m'$, $\bar{4}'2'm$
$\dfrac{2}{m}$	$\dfrac{2}{m'}$, $\dfrac{2'}{m}$, $\dfrac{2'}{m'}$
$\dfrac{4}{m}$	$\dfrac{4}{m'}$, $\dfrac{4'}{m}$, $\dfrac{4'}{m'}$
$\dfrac{6}{m}$	$\dfrac{6}{m'}$, $\dfrac{6'}{m}$, $\dfrac{6'}{m'}$
$\dfrac{2}{m}\dfrac{2}{m}\dfrac{2}{m}$	$\dfrac{2}{m'}\dfrac{2}{m'}\dfrac{2}{m'}$, $\dfrac{2}{m'}\dfrac{2'}{m}\dfrac{2'}{m'}$, $\dfrac{2}{m}\dfrac{2'}{m'}\dfrac{2'}{m}$
$\dfrac{4}{m}\dfrac{2}{m}\dfrac{2}{m}$	$\dfrac{4}{m'}\dfrac{2}{m'}\dfrac{2}{m'}$, $\dfrac{4}{m'}\dfrac{2'}{m}\dfrac{2'}{m}$, $\dfrac{4}{m}\dfrac{2'}{m'}\dfrac{2'}{m'}$, $\dfrac{4'}{m}\dfrac{2}{m'}\dfrac{2'}{m'}$, $\dfrac{4'}{m'}\dfrac{2}{m'}\dfrac{2}{m}$
$\dfrac{6}{m}\dfrac{2}{m}\dfrac{2}{m}$	$\dfrac{6}{m'}\dfrac{2}{m'}\dfrac{2}{m'}$, $\dfrac{6}{m'}\dfrac{2'}{m}\dfrac{2'}{m}$, $\dfrac{6}{m}\dfrac{2'}{m'}\dfrac{2'}{m'}$, $\dfrac{6'}{m}\dfrac{2}{m'}\dfrac{2'}{m}$, $\dfrac{6'}{m}\dfrac{2}{m'}\dfrac{2'}{m'}$
$\dfrac{2}{m}\bar{3}$	$\dfrac{2}{m'}\bar{3}'$
432	$4'32'$
$\bar{4}3m$	$\bar{4}'3m'$
$\dfrac{4}{m}\bar{3}\dfrac{2}{m}$	$\dfrac{4}{m'}\bar{3}'\dfrac{2}{m'}$, $\dfrac{4'}{m'}\bar{3}'\dfrac{2'}{m}$, $\dfrac{4'}{m}\bar{3}\dfrac{2'}{m'}$

Chapter 5

Space Lattices

5.1 TWO-DIMENSIONAL LATTICES

Crystals are objects of common experience. The diamond, for instance, is a single crystal built up from a periodic array of carbon atoms. Any piece of metal we handle consists of numerous single crystals of random size, shape and orientation (Fig. 5.1). A two-dimensional crystal model is displayed in Fig. 5.2. This exhibits two distinct features: the recurring motif X, which

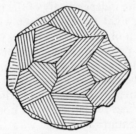

Fig. 5.1—Polycrystalline aggregate making up bulk metal.

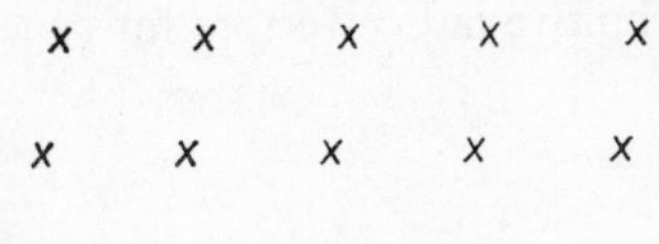

$$X \quad X \quad X \quad X \quad X$$

$$X \quad X \quad X \quad X \quad X$$

$$X \quad X \quad X \quad X \quad X$$

Fig. 5.2—Two-dimensional crystal model.

indicates some grouping of atoms or ions, and the way X is repeated throughout space. The motif may comprise an enormous number of atoms, as in some organic crystals, or it may comprise only one or two, as in most metals. Further consideration of X, and its relation to the symmetry patterns of Part I, will be given in Chapter 6. Here we are primarily concerned with translational aspects.

Fig. 5.3—Reference lattice of two-dimensional crystal.

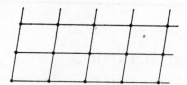

Fig. 5.4—Two-dimensional space lattice.

The repetition of X through space is conveniently defined by means of a reference lattice (Fig. 5.3). This is a mathematical construct, which can be introduced and studied quite independently of the motif structure. It consists of two sets of parallel straight lines, each set with its own characteristic spacing (Fig. 5.4). The point of intersection of any two lines is a lattice point; the join of any two lattice points is a lattice vector. Two non-parallel lattice vectors, emanating from a lattice point, serve to define a parallelogram termed a unit cell. This is said to be primitive if it contains no interior lattice points, being otherwise non-primitive. Primitive cells are the more fundamental since such a cell immediately defines the lattice, but a non-primitive cell can sometimes exhibit symmetry features not otherwise apparent (Fig. 5.5). The original intersecting lines define an immediately obvious primitive unit cell, but this has no more significance than any other primitive cell. To put the matter somewhat differently, any method of constructing the lattice amounts to choosing a pair of primitive lattice vectors for generating lattice points.

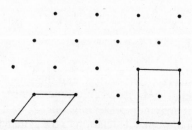

Fig. 5.5—Primitive and non-primitive unit cells. The latter exhibits rectangular symmetry.

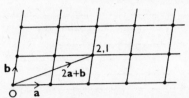

Fig. 5.6—Two-dimensional lattice generated by **a** and **b**, showing the lattice
vector $2\mathbf{a} + \mathbf{b}$ terminating in the lattice point [2,1].

We now introduce coordinates. An arbitrary lattice point is chosen as
origin, and a pair of primitive lattice vectors **a**, **b** chosen as generating vectors
(Fig. 5.6). Any two integers, x, y—positive, negative, or zero—then define
another lattice vector $x\mathbf{a} + y\mathbf{b}$ terminating in a lattice point denoted $[x, y]$. The
vectors **a**, **b** emanate from [0, 0] and terminate in [1, 0], [0, 1] respectively.
Note that there is just one lattice point per primitive unit cell, so that all such
cells, whatever their orientation, have the same area $|\mathbf{a} \wedge \mathbf{b}|$. Straight lines
through the origin may be rational or irrational, meaning that they either do
or do not pass through other lattice points. A rational line has the equation

$$hx + ky = 0 \tag{1}$$

where h, k are relatively prime integers and $[x, y]$ is a lattice point on the line.
All integer solutions of equation (1) may be found without explicitly specifying h and k, being evidently

$$0, 0; \quad k, \overline{h}; \quad 2k, 2\overline{h}; \quad \ldots; \quad nk, n\overline{h}; \ldots \tag{2}$$

and they define a succession of lattice points characterised by the repeat
vector $\mathbf{R} = k\mathbf{a} - h\mathbf{b}$. The equation

$$hx + ky = 1 \tag{3}$$

defines a rational line parallel to (1) and characterised by the same repeat
vector **R**. This follows since Euclid's algorithm assures at least one integer
solution x_0, y_0 of (3), upon which may be superposed any solution of (1),
thereby yielding an infinite number of solutions of the form

$$x_0 + nk, y_0 - nh; \quad n = 0, \pm 1, \pm 2, \ldots \tag{4}$$

These two lines bear a simple geometrical relationship, since the second can be
generated from the first by a rigid-body translation made up of a normal
component **d** and a tangential component **t** (Fig. 5.7). Similarly proceeding,
we arrive at the set of parallel rational lines

$$hx + ky = 0, \pm 1, \pm 2, \ldots \tag{5}$$

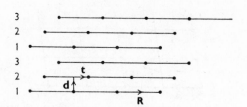

Fig. 5.7—Model of repeat vector **R**, interplanar spacing vector **d**, and shift vector **t**. Note that $\mathbf{R} = 3\mathbf{t}$, showing that the stacking pattern is ......123123......

each characterised by **R** and generated from its predecessor by $\mathbf{d}+\mathbf{t}$. The stacking properties of the set are determined by $\mathbf{t}$: if $\mathbf{R} = p\mathbf{t}$, the number p may either be rational, in which case a definite stacking pattern exists, or p may be irrational, in which case no stacking pattern exists. It is intuitively clear that this set of lines includes every lattice point, as also follows formally from the result

$$|\mathbf{R} \wedge (\mathbf{d}+\mathbf{t})| = |(k\mathbf{a} - h\mathbf{b}) \wedge (x_0\mathbf{a}+y_0\mathbf{b})| = (hx_0 + ky_0)|\mathbf{a} \wedge \mathbf{b}| = |\mathbf{a} \wedge \mathbf{b}|,$$

on bearing in mind that an integer n can always be chosen so that

$$\mathbf{d}+\mathbf{t} = (x_0\mathbf{a} + y_0\mathbf{b}) + n(k\mathbf{a} - h\mathbf{b}).$$

5.2 THREE-DIMENSIONAL LATTICES

The extension of the preceding ideas to three dimensions is reasonably straightforward. Given three sets of parallel planes, each set with its own characteristic interplanar spacing, the intersection of any three planes is a lattice point, and the join of any two lattice points is a lattice vector. Three non-coplanar lattice vectors serve to define a parallelepiped, termed a unit cell, which may be either primitive or non-primitive, and all primitive cells, whatever their orientation, have the same volume. We introduce coordinates by choosing an arbitrary lattice point as origin, and a triplet of primitive lattice vectors **a**, **b**, **c** as generating vectors. Any three integers x, y, z then define another lattice vector $x\mathbf{a} + y\mathbf{b} + z\mathbf{c}$ terminating in the lattice point denoted $[x, y, z]$. The vectors **a**, **b**, **c** emanate from $[0, 0, 0]$ and terminate in $[1, 0, 0]$, $[0, 1, 0]$, $[0, 0, 1]$ respectively. Planes through the origin may be rational or irrational. Both play a role in physical applications, but only the former will be considered here.

A rational plane, usually termed a lattice plane since it is a two-dimensional lattice, has the equation

$$hx + ky + lz = 0, \tag{6}$$

where h, k, l are three relatively prime integers compactly written (hkl) and known as Miller indices. Integer solutions of (6) define lattice points lying in the plane, and the problem immediately arises of determining these and ordering them into a geometrically significant pattern. Obvious integer solutions of (6) are

$$0, 0, 0; \quad l, 0, \overline{h}; \quad 0, l, \overline{k}; \quad k, \overline{h}, 0 \tag{7}$$

and further solutions are generated by the superposition process

$$\alpha[l, 0, \overline{h}] + \beta[0, l, \overline{k}] = [\alpha l, \beta l, \overline{\alpha h + \beta k}], \tag{8}$$

where α, β are arbitrary integers. However, not all the possible integer solutions of (6) are generated by (8) since, for instance, no admissible choice of α, β can satisfy

$$[\alpha l, \beta l, \overline{\alpha h + \beta k}] = [k, \overline{h}, 0]$$

unless $l = 1$ (in which case $\alpha = k$, $\beta = -h$). Geometrically expressed, we have defined a pair of lattice vectors

$$\mathbf{B} = l\mathbf{a} - h\mathbf{c}, \quad \mathbf{A} = l\mathbf{b} - k\mathbf{c} \tag{9}$$

in the plane, and hence a unit cell in the plane, which is primitive only if $l = 1$. To locate its interior points, construct the set of $l - 1$ lines

$$x = 1, 2, \ldots, l-1; \quad hx + ky + lz = 0 \tag{10}$$

parallel to the sides $x = 0, l$; $hx + ky + lz = 0$ of the cell, and also the set of lines

$$y = 1, 2, \ldots, l-1; \quad hx + ky + lz = 0 \tag{11}$$

parallel to the other two sides $y = 0, l$; $hx + ky + lz = 0$ (Fig. 5.8). Each line of (10) contains one interior lattice point. To prove this, note that $x = 1$; $hx + ky + lz = 0$ imply $h + ky + lz = 0$ which, unless k and l have a common

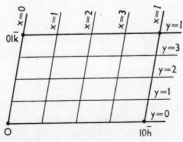

Fig. 5.8—Procedure for location of interior points within unit cell.

factor, always admits an integer solution y_0, z_0 and hence an infinite number of solutions

$$1, y_0 + nl, z_0 - nk; \quad n = 0, \pm1, \pm2, \ldots$$

one of which satisfies the condition $0 < y_0 + nl < l$ distinguishing it as an interior lattice point. A similar analysis holds for $x = 2, \ldots, l - 1$. We see, therefore, that there exist a maximum of $l - 1$ interior lattice points altogether. Figure 5.9 shows this procedure applied to the (295) plane of a primitive cubic lattice. Since h and k have the same status as l, it follows that a maximum of $h - 1$ interior lattice points exist inside the cell **B**, **C** $= k\mathbf{a} - h\mathbf{b}$, and $k - 1$ inside the cell **A**, **C**.

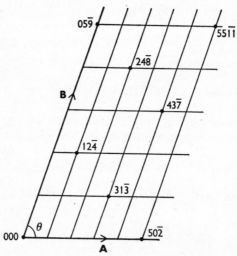

Fig. 5.9—Primitive cubic lattice plane (295) drawn to scale. For the sake of compactness, commas have been omitted from the coordinate numbers.

The equation $hx + ky + lz = 1$ defines a rational plane parallel to (6) and characterised by the same arrangement of lattice points. It is generated from (6) by a rigid-body translation $\mathbf{d} + \mathbf{t}$ made up of a normal component $\mathbf{d}$ (interplanar spacing vector) and a tangential component $\mathbf{t}$ (stacking vector). Similarly proceeding, we generate the set of parallel lattice planes

$$hx + ky + lz = \pm1, \pm2, \ldots$$

which, between them, contain all the lattice points. The vector $\mathbf{d}$ may be calculated explicitly in terms of $\mathbf{a}$, $\mathbf{b}$, $\mathbf{c}$; h, k, l, and a formula is given in Appendix 5. The calculation of $\mathbf{t}$ presents a much more difficult problem, which only admits a solution subject to the same limitations as that of the lattice arrangement problem, i.e. it must be solved by an algorithm except when h, k or $l = 1$. Details are given in Appendix 5.

5.3 UNIT CELLS

To say that a unit cell is defined by **a**, **b**, **c** means that we know the lengths $|\mathbf{a}| = a$, $|\mathbf{b}| = b$, $|\mathbf{c}| = c$, and the angles α, β, γ between **b**,**c**; **c**,**a**; **a**,**b** respectively. In terms of these six quantities, we have

$$V = \mathbf{a}.\mathbf{bc} = \begin{vmatrix} a^2 & \mathbf{a}.\mathbf{b} & \mathbf{a}.\mathbf{c} \\ \mathbf{b}.\mathbf{a} & b^2 & \mathbf{b}.\mathbf{c} \\ \mathbf{c}.\mathbf{a} & \mathbf{c}.\mathbf{b} & c^2 \end{vmatrix}^{\frac{1}{2}} \tag{12}$$

for the cell volume, and

$$r = |x\mathbf{a} + y\mathbf{b} + z\mathbf{c}| = \{(x\mathbf{a} + y\mathbf{b} + z\mathbf{c}).(x\mathbf{a} + y\mathbf{b} + z\mathbf{c})\}^{\frac{1}{2}} \tag{13}$$

for the length of a lattice vector. Utilising digital computer techniques, it would be entirely feasible to calculate r systematically as a function of the integers x, y, z and hence to order all the lattice vectors into a monotonic sequence depending on length. The three non-coplanar lattice vectors **A**, **B**, **C** of shortest length are not, of course, necessarily identical with **a**, **b**, **c**. Given two apparently different lattices generated by $\mathbf{a}_1$, $\mathbf{b}_1$, $\mathbf{c}_1$ and $\mathbf{a}_2$, $\mathbf{b}_2$, $\mathbf{c}_2$ respectively, how do we know whether or not they are identical, i.e. whether or not one can be brought into complete coincidence with the other by a rigid-body translation and rotation? The most straightforward procedure would be to construct the **A**, **B**, **C** cell for each case thus providing a significant basis for comparison.*

The point-group symmetry of space lattices will be studied in the next two chapters. Two-dimensional lattices and crystal models may be depicted directly. However, three-dimensional lattices are best understood in terms of the unit cells which generate them. This theory paves the way for an introductory analysis of three-dimensional crystals.

*Further details may be found in the papers of Bevis [20], Nicholas [21] and Gruber [22].

Chapter 6

The Seven Crystal Systems

6.1 ROTATIONAL SYMMETRY OF TWO-DIMENSIONAL LATTICES

The arbitrary two-dimensional lattice (Fig. 5.4) has a 2-fold symmetry axis perpendicular to its plane, passing through any lattice point 0. Of course 0 also functions as a symmetry centre, but this adds nothing to the 2-fold axis. Next comes the rectangular net (Fig. 6.1), which has a 2-fold symmetry axis

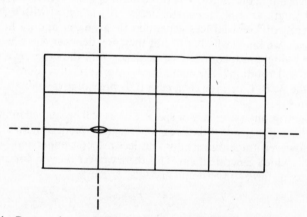

Fig. 6.1—Rectangular net indicating a principal 2-fold axis and traces (dotted lines) of axial mirror planes.

perpendicular to its plane plus two axial symmetry planes as indicated. A more symmetric net than this is the square net (Fig. 6.2), which has a 4-fold symmetry axis perpendicular to its plane plus four axial symmetry planes. The most symmetric possible net is the hexagonal net (Fig. 6.3), which has a 6-fold symmetry axis perpendicular to its plane plus six axial symmetry planes. This may be generated by a 120° rhombus, and it therefore ranks as a two-dimensional lattice in the sense of section 5.1.

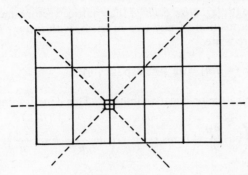

Fig. 6.2—Square net indicating a 4-fold axis and traces (dotted lines) of axial
mirror planes.

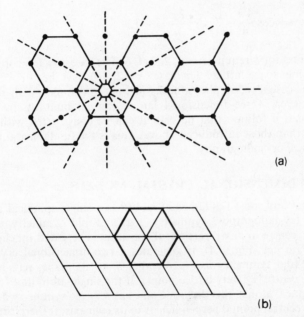

(a)

(b)

Fig. 6.3—(a) Hexagonal net indicating a 6-fold axis and traces (dotted lines) of
axial mirror planes. (b) Generation of hexagonal net by a rhombus of angle 120°.

We now go over the preceding ground using the algebra of complex
numbers. Let z be the complex number representing a lattice vector emanating from 0. If an n-fold symmetry axis passes through 0, perpendicular
to the lattice plane, it generates n independent lattice vectors

$$z, z \exp(2\pi i/n), \quad z \exp(4\pi i/n), \ldots$$

from z, all of length $|z|$. Now each of these lattice vectors must be expressible as a linear superposition of any two of them, with integer coefficients, and we may therefore write

$$z \exp (4\pi i/n) = az + bz \exp (2\pi i/n)$$

where a, b are positive or negative integers. This equality implies

$$a = \left(\cos \frac{4\pi}{n} - b \cos \frac{2\pi}{n} \right) + i \left(\sin \frac{4\pi}{n} - b \sin \frac{2\pi}{n} \right),$$

whence

$$\sin \frac{4\pi}{n} - b \sin \frac{2\pi}{n} = 0, \quad \text{i.e. } b = 2 \cos \frac{2\pi}{n},$$

$$a = \cos \frac{4\pi}{n} - 2 \cos^2 \frac{2\pi}{n} = -1.$$

Integer values for b require $n = 1, 2, 3, 4, 6$. The case $n = 1$ is consistent with every possible space lattice. The cases $n = 2, 4, 6$ have been covered above. Finally, the case $n = 3$ implies a hexagonal net in common with $n = 6$.

Since every three-dimensional lattice can be built up from parallel lattice planes, it follows that no crystal can be constructed with symmetry axes other than those introduced above. This is the fundamental theorem of mathematical crystallography.

6.2 TWO-DIMENSIONAL CRYSTAL MODELS

The simplest and most fundamental motif* is a single spherical atom. This has every crystallographic symmetry, e.g. a 2-fold symmetry axis passes through its centre in any direction. If the atom is repeated indefinitely on a parallelogram net (Fig. 6.4), it generates a two-dimensional crystal model having a 2-fold symmetry axis perpendicular to its plane, passing through any atomic centre (i.e. any lattice point). If the single-atom motif is replaced by two identical atoms, this diatomic unit has a 2-fold symmetry axis passing through its central point 0 perpendicular to its own axis. It therefore generates

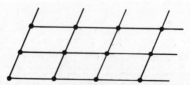

Fig. 6.4—Parallelogram net with a single atom located at each lattice point.

* See section 5.1.

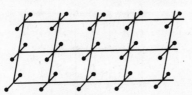

Fig. 6.5—Diatomic motif pattern (consisting of two identical A-atoms) on a
parallelogram net.

a crystal with a 2-fold symmetry axis (Fig. 6.5). If, however, the two atoms
are distinct (A, B), this 2-fold axis reduces to a 1-fold axis and the resulting
crystal has no particular symmetry (Fig. 6.6). These macroscopic conclusions
are intuitively clear from the diagrams, but they are also covered by the
general theory of Chap. 9.

The single spherical atom (A-atom) has a 4-fold symmetry axis passing
through its centre in any direction. This motif may be expanded without
impairing its symmetry, by introducing four additional atoms (B-atoms) as
displayed in Fig. 6.7. If X is now translated over a square net, it generates a
crystal characterised by a 4-fold principal symmetry axis passing through
any A-atom (i.e. any lattice point). A non-trivial variant of Fig. 6.7 appears
in Fig. 6.8, where each B-atom lies at the centre of a square delineated by
A-atoms. The generating motif is of the same type and symmetry as its
predecessor but overlap now occurs since each B-atom belongs to four
neighbouring motifs. In effect, therefore, only one-quarter of each B-atom

Fig. 6.6—Diatomic motif pattern consisting of an A-atom (marked ●) and a B-atom
(marked .) on a parallelogram net.

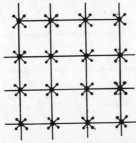

Fig. 6.7—Square net with motif pattern consisting of an A-atom at each lattice
point surrounded symmetrically by four B-atoms.

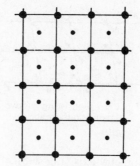

Fig. 6.8—Previous motif pattern with each B-atom at the centre of a square
delineated by A-atoms.

belongs to its central *A*-atom, which means one *B*-atom per *A*-atom on the
average—compared with four *B*-atoms per *A*-atom in Fig. 6.7. Clearly there
exists an alternative unit of repetition, which consists of one *A*-atom and a
neighbouring *B*-atom. This unit generates the crystal without overlap, but it
lacks the rotational symmetry inherent in the macroscopic crystal.

The crystal model depicted in Fig. 6.9 is generated on a hexagonal net
by a triangular motif, i.e. the *A*-atom at each lattice point surrounded
symmetrically by three *B*-atoms, so exhibiting a 3-fold principal symmetry
axis. In Fig. 6.10, each *B*-atom lies at the centre of an equilateral triangle
delineated by *A*-atoms, so providing an alternative unit of repetition remi-
niscent of Fig. 6.8. This model will help in understanding the close-packed
hexagonal structure discussed in section 6.5. Hexagonal nets are, of course,
also compatible with a 6-fold principal symmetry axis, which is ensured by
the introduction of a regular hexagonal motif as depicted in Fig. 6.11.

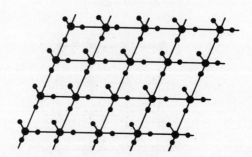

Fig. 6.9—Hexagonal net with triangular motif consisting of an A-atom at each
lattice point surrounded symmetrically by three B-atoms.

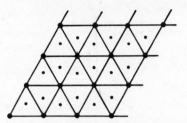

Fig. 6.10—Previous motif pattern with each B-atom at the centre of an equilateral triangle delineated by A-atoms.

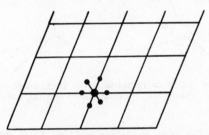

Fig. 6.11—Hexagonal motif pattern on a hexagonal net.

6.3 THE SEVEN CRYSTAL SYSTEMS

Three-dimensional space lattices are generated by parallelepipeds termed unit cells. No cell can be more symmetric than the cube. Extending or contracting one of its edges produces the tetragonal cell. Extending or contracting a second edge produces the orthorhombic cell. So far all the cell edges have remained mutually orthogonal. Changing one of the angles from $\pi/2$ produces the monoclinic cell. Changing a second angle from $\pi/2$ produces the triclinic cell, which has no special symmetry features. This cell must therefore be specified by giving all three edge lengths $\mathbf{a}$, $\mathbf{b}$, $\mathbf{c}$ and corresponding angles α, β, γ mentioned in section 5.4. There exist two further cells which do not fit directly into the preceding scheme. First, extending or contracting a cube uniformaly along one of its diagonals produces a rhombohedron, the three-dimensional analogue of a rhombus. Secondly, changing the angle $\gamma(= \pi/2)$ of the tetragonal cell from $\pi/2$ to $2\pi/3$ produces a hexagonal cell, so called since this cell generates the primitive hexagonal lattice. We prove in Chapter 7 that the rhombohedral cell is equivalent to a double-centred hexagonal cell. However, it often appears convenient to treat this as a distinct system. Accordingly the unit cells may be classified into seven standard systems depending on their symmetry properties.

Three concurrent edges of each cell are exhibited in Fig. 6.12, these being identified as the generating lattice vectors $\mathbf{a}$, $\mathbf{b}$, $\mathbf{c}$. We often use the component representations

$$\mathbf{a} = <1, 0, 0>, \qquad \mathbf{b} = <0, 1, 0>, \qquad \mathbf{c} = <0, 0, 1>$$

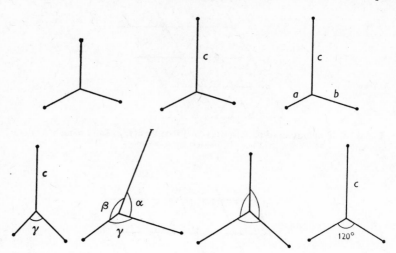

Fig. 6.12—Edges of the seven primitive lattice cells: cubic, tetragonal, ortho-rhombic, monoclinic, triclinic, rhombohedral, hexagonal.

enabling us to write the edge directions as [100], [010], [001] and the corresponding cell faces as (100), (010), (001). The seven full cells are exhibited in the first and final columns of Fig. 6.13, p. 000, together with the seven independent non-primitive cells (see Chapter 7) which share their symmetry properties. These make up the fourteen Bravais cells of classical crystallography.

When unit cells are stacked together to form a space lattice, this has a definite point-group symmetry with respect to any lattice point 0 as the fixed point. Thus, starting with the primitive cubic lattice, the three cell edges through 0 clearly constitute 4-fold symmetry axes for the lattice. Also, the four cell diagonals through 0 constitute 3-fold symmetry axes for the lattice. These intersecting symmetry axes provide the point-group symmetry *432* (Chapter 3), which may be extended at once to $\frac{4}{m} \bar{3} \frac{2}{m}$ since 0 functions as a symmetry centre. This is the holohedral cubic symmetry, meaning that it covers all the lower cubic symmetries *23*, $\frac{2}{m} \bar{3}$, *$\bar{4}$3m*, *432*. Proceeding in similar fashion for the primitive tetragonal, orthorhombic and monoclinic lattices, we obtain the results displayed in Table 9.1. The triclinic lattice does not possess symmetry axes or planes, but 0 functions as a symmetry centre so providing the holohedral symmetry *$\bar{1}$* for this system.

The rhombohedral space lattice has a 3-fold symmetry axis coinciding with an extended or contracted cube diagonal. Also three symmetry planes pass through this diagonal, so providing the symmetry *3m* which may be extended to the holohedral symmetry $\bar{3} \frac{2}{m}$. This covers the lower trigonal

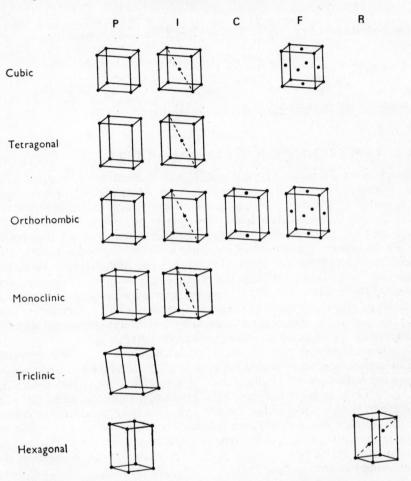

P   I   C   F   R

Cubic

Tetragonal

Orthorhombic

Monoclinic

Triclinic

Hexagonal

Fig. 6.13—The fourteen Bravais unit cells. Primitive cells appear under the column P; body-centred cells appear under I; end-centred cells appear under C; and face-centred cells under F. The rhombohedral cell of Fig. 6.12 has been replaced by the equivalent non-primitive hexagonal cell appearing under R.

symmetries $3, \bar{3}, 32, 3m$ but not $\frac{3}{m}, \frac{3}{m}m2$ which are not subsymmetries of $\bar{3}\frac{2}{m}$.
In section 7.4 we identify the rhombohedral lattice as a doubly-centred hexagonal lattice, which allows an alternative insight into its symmetry properties. There remains the primitive hexagonal lattice, characterised by a 6-fold symmetry axis coinciding with the principal cell edge and by 2-fold symmetry axes coinciding with the secondary edges. These intersecting axes provide the point-group symmetry *622*, which may be extended to the

holohedral symmetry $\dfrac{6}{m}\dfrac{2}{m}\dfrac{2}{m}$. This covers all the lower symmetries

$$\frac{3}{m}, \frac{3}{m}m2; \; 6, \frac{6}{m}, 6mm, 622$$

as well as the rhombohedral symmetries $\bar{3}\dfrac{2}{m}$, etc.

6.4 THREE-DIMENSIONAL CRYSTAL MODELS

Simple three-dimensional crystal models may be constructed by introducing a suitable motif pattern at each lattice point. More precisely, we envisage the pattern as having a point-group symmetry $\{G\}$ with respect to 0, and we assume that the space lattice has a point-group symmetry of at least $\{G\}$ with respect to 0. If the pattern is repeated by translation at each lattice point, then the whole (infinite) crystal structure has a point-group symmetry $\{G\}$ with respect to 0, as may be proved either geometrically (Hilton) or by the methods of Chapter 9. This is the crystal's microscopic point-group symmetry. The macroscopic crystal may be regarded as a continuum without structure. However it has a point-group symmetry which can be determined by physical and geometrical methods, and which is clearly identical with the microscopic point-group symmetry in this particular model.

  The atoms may possibly lie at special positions within the unit cell. This causes them to be shared between motif patterns centred upon neighbouring lattice points, as illustrated for some two-dimensional models in section 6.2. If so, we may choose an alternative, asymmetric, motif made up of atoms lying entirely within the unit cell. This motif generates the crystal structure most economically, but it lacks any symmetry features related to the macroscopic crystal. Either type of motif may be chosen as the starting point of a geometrical analysis, as will be shown by an example in the following section.

6.5 CLOSE-PACKED HEXAGONAL STRUCTURE

This structure, adopted by several important metals, may be formally generated by a hexagonal unit cell containing two identical atoms located at $[0, 0, 0]$, $\left[\dfrac{2}{3}, \dfrac{1}{3}, \dfrac{1}{2}\right]$—see Fig. 6.14. An alternative approach is to picture the atoms as equal, rigid, spheres arranged in close-packed layers. In any layer the centres form a hexagonal net (Fig. 6.15), and one of these is chosen to be the initial (basal) net comprising all the lattice points

$$[x, y, 0]; \; x, y = 0, \pm 1, \pm 2, \ldots$$

These co-ordinates refer to a triplet of generating lattice vectors which define a primitive hexagonal cell standing on the net. If a second layer of spheres is

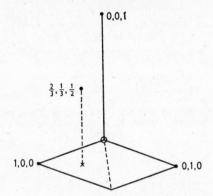

Fig. 6.14—Hexagonal cell indicating location of interior atom for the close-packed hexagonal structure.

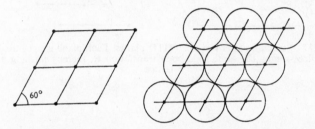

Fig. 6.15—Arrangement of atoms in the hexagonal (0001) plane, or equivalently in the face-centred cubic (111) plane, this being realized by the centres of close-packed equal rigid spheres.

deposited on the first, as close-packed as possible, their centres define a second hexagonal net which projects upon the first as indicated in Fig. 6.16. This net may accordingly be generated from its predecessor by a rigid-body translation vector $<\frac{2}{3}, \frac{1}{3}, \frac{1}{2}>$, so accounting for the interior atom within each cell. The third net may be generated from its predecessor in two distinct, though equally admissible, ways:

either (a) by a rigid-body translation vector $<\frac{2}{3}, \frac{1}{3}, \frac{1}{2}>$, etc., so producing the face-centred cubic stacking pattern ... 123123 ... (Fig. 6.17), further discussed in Chapter 7;

or (b) by a rigid-body translation vector $<-\frac{2}{3}, -\frac{1}{3}, \frac{1}{2}>$, so projecting directly upon the basal net and producing the close-packed hexagonal stacking pattern ... 121212 ... (Fig. 6.18), further discussed in Chapter 7.

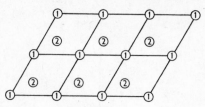

Fig. 6.16—Stacking pattern of c.p.h. (0001) planes. These planes are generated by alternate rigid-body translations **R, R′** as shown, so that a space lattice is not formed.

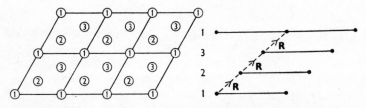

Fig. 6.17—Stacking pattern of f.c.c. (111) planes. Each plane is generated from its predecessor by the same rigid-body translation **R**, thereby forming a space lattice.

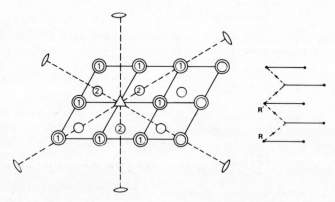

Fig. 6.18—Stacking pattern of c.p.h. configuration, which is the same as in Fig. 6.16 except that ⓵ replaces ① to mark the projection of plane 3 on plane 1.

A 3-fold symmetry axis evidently passes through any ⓵-point of Fig. 6.18 normal to the net. Also, three symmetry planes pass through this axis and its neighbouring ②-points. Finally, the net itself functions as a symmetry plane by virtue of the stacking pattern . . . 121212 . . . . These symmetry features imply the point-group symmetry $\frac{3}{m}m2$ centred about any

①-point. Thus, for instance, we may generate the six equivalent ②-points

$$\left[\frac{2}{3}, \frac{1}{3}, \frac{1}{2}\right], \left[\frac{\bar{1}}{3}, \frac{1}{3}, \frac{1}{2}\right], \left[\frac{\bar{1}}{3}, \frac{\bar{2}}{3}, \frac{1}{2}\right]; \left[\frac{2}{3}, \frac{1}{3}, \frac{\bar{1}}{2}\right], \left[\frac{\bar{1}}{3}, \frac{1}{3}, \frac{\bar{1}}{2}\right], \left[\frac{\bar{1}}{3}, \frac{\bar{2}}{3}, \frac{\bar{1}}{2}\right] \qquad (1)$$

from the ②-point $\left[\frac{2}{3}, \frac{1}{3}, \frac{1}{2}\right]$, keeping the ①-point $[0, 0, 0]$ fixed. All these points, including $[0, 0, 0]$, are of course occupied by atoms, so providing a seven-atom motif pattern centred about the lattice point $[0, 0, 0]$. An identical pattern surrounds every lattice point. However, the neighbouring patterns overlap, e.g. the atom at $\left[\frac{2}{3}, \frac{1}{3}, \frac{1}{2}\right]$ is shared by the six patterns centred respectively on the lattice points

$$[0, 0, 0], [1, 0, 0], [1, 1, 0]; \qquad [0, 0, 1], [1, 0, 1], [1, 1, 1] \qquad (2)$$

Accordingly the six atoms at (1) contribute, on the average, only one complete atom per lattice point, so reconciling this seven-atom symmetric motif with the two-atom asymmetric motif introduced in Fig. 6.14.

Finally, we note that the primitive hexagonal lattice has a point-group symmetry of at least $\frac{3}{m}m2$. Therefore the crystal as a whole has this symmetry, at both the microscopic and macroscopic levels.

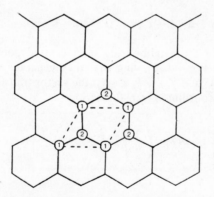

Fig. 6.19—Generation of hexagonal honeycomb pattern by a rhombus of ① -points containing a ② -point. See Fig. 6.16 and footnote p. 85.

Chapter 7

Non-primitive Units Cells

7.1 BODY-CENTRED CELLS

A non-primitive cell is defined by three non-coplanar lattice vectors $\mathbf{a}$, $\mathbf{b}$, $\mathbf{c}$ as before, but the superposition process $x\mathbf{a} + y\mathbf{b} + z\mathbf{c}$ now generates only corner lattice points. Body-centred lattice points are generated by the superposition

$$(x + \tfrac{1}{2})\mathbf{a} + (y + \tfrac{1}{2})\mathbf{b} + (z + \tfrac{1}{2})\mathbf{c},$$

so that every unit cell contains an interior lattice point located at $\tfrac{1}{2}(\mathbf{a} + \mathbf{b} + \mathbf{c})$ relative to the cell edges. However, it requires proof that the combined array of points so produced, i.e. those having coordinates $[x, y, z]$ and those having coordinates $[x + \tfrac{1}{2}, y + \tfrac{1}{2}, z + \tfrac{1}{2}]$, constitutes a space lattice. This proof is achieved by choosing three vectors

$$\mathbf{A} = \tfrac{1}{2}(-\mathbf{a} + \mathbf{b} + \mathbf{c}), \qquad \mathbf{B} = \tfrac{1}{2}(\mathbf{a} - \mathbf{b} + \mathbf{c}), \qquad \mathbf{C} = \tfrac{1}{2}(\mathbf{a} + \mathbf{b} - \mathbf{c}) \quad (1)$$

joining $[0, 0, 0]$ to the neighbouring body-centred points $[\bar{\tfrac{1}{2}}, \tfrac{1}{2}, \tfrac{1}{2}]$, $[\tfrac{1}{2}, \bar{\tfrac{1}{2}}, \tfrac{1}{2}]$, $[\tfrac{1}{2}, \tfrac{1}{2}, \bar{\tfrac{1}{2}}]$ respectively, and showing that they generate a space lattice which embraces $[x, y, z]$, $[x + \tfrac{1}{2}, y + \tfrac{1}{2}, z + \tfrac{1}{2}]$: the vector transformation (1) induces a corresponding coordinate transformation

$$\begin{pmatrix} X \\ Y \\ Z \end{pmatrix} = (P/I) \begin{pmatrix} \chi \\ \eta \\ \zeta \end{pmatrix} = \begin{pmatrix} \eta + \zeta \\ \zeta + \chi \\ \chi + \eta \end{pmatrix}, \tag{2}$$

where

$$(P/I) = \begin{pmatrix} 0 & 1 & 1 \\ 1 & 0 & 1 \\ 1 & 1 & 0 \end{pmatrix}, \qquad (I/P) = (P/I)^{-1} = \tfrac{1}{2} \begin{pmatrix} \bar{1} & 1 & 1 \\ 1 & \bar{1} & 1 \\ 1 & 1 & \bar{1} \end{pmatrix}$$

$$[\chi, \eta, \zeta] = [x, y, z] \text{ or } [x + \tfrac{1}{2}, y + \tfrac{1}{2}, z + \tfrac{1}{2}]$$

and $[X, Y, Z]$ are coordinates referred to the primitive unit cell (1), these being integers for both choices of χ, η, ζ as can be seen from (2).

Our argument shows that the combined arrays are included within the space lattice generated by (1), but it does not prove the complete identification

$$[x, y, z], \quad [x+\tfrac{1}{2}, y+\tfrac{1}{2}, z+\tfrac{1}{2}] \rightleftarrows [X, Y, Z].$$

This follows by noting that

$$(I/P)\begin{pmatrix} X \\ Y \\ Z \end{pmatrix} = \tfrac{1}{2}\begin{pmatrix} -X+Y+Z \\ X-Y+Z \\ X+Y-Z \end{pmatrix} = \begin{pmatrix} x \\ y \\ z \end{pmatrix} \text{ or } \begin{pmatrix} x+\tfrac{1}{2} \\ y+\tfrac{1}{2} \\ z+\tfrac{1}{2} \end{pmatrix}$$

for any triplet of integers X, Y, Z. The most important application of (1) is to the body-centred cubic lattice, in which case $\mathbf{A}, \mathbf{B}, \mathbf{C}$ define a primitive rhombohedral cell of angle $\cos^{-1}(-\tfrac{1}{3})$.

7.2 FACE-CENTRED CELLS

Face-centred lattice points are of the form

$$[x+\tfrac{1}{2}, y+\tfrac{1}{2}, z], \quad [x, y+\tfrac{1}{2}, z+\tfrac{1}{2}], \quad [x+\tfrac{1}{2}, y, z+\tfrac{1}{2}] \tag{3}$$

but again it requires proof that these combined arrays, together with $[x, y, z]$, constitute a space lattice. This is achieved by choosing three vectors

$$\mathbf{A}=\tfrac{1}{2}(\mathbf{b}+\mathbf{c}), \quad \mathbf{B}=\tfrac{1}{2}(\mathbf{c}+\mathbf{a}), \quad \mathbf{C}=\tfrac{1}{2}(\mathbf{a}+\mathbf{b}) \tag{4}$$

joining $[0, 0, 0]$ to the neighbouring face-centred points $[0, \tfrac{1}{2}, \tfrac{1}{2}]$, $[\tfrac{1}{2}, 0, \tfrac{1}{2}]$, $[\tfrac{1}{2}, \tfrac{1}{2}, 0]$ respectively, and showing that they generate a space lattice which embraces the four arrays. Thus, corresponding to (2), we have

$$\begin{pmatrix} X \\ Y \\ Z \end{pmatrix} = (P/F)\begin{pmatrix} x \\ \eta \\ \zeta \end{pmatrix} = \begin{pmatrix} -x+\eta+\zeta \\ x-\eta+\zeta \\ x+\eta-\zeta \end{pmatrix}, \tag{5}$$

where

$$(P/F)=\begin{pmatrix} \bar{1} & 1 & 1 \\ 1 & \bar{1} & 1 \\ 1 & 1 & \bar{1} \end{pmatrix}, \quad (F/P)=\tfrac{1}{2}\begin{pmatrix} 0 & 1 & 1 \\ 1 & 0 & 1 \\ 1 & 1 & 0 \end{pmatrix}$$

$[\chi, \eta, \zeta] = [x, y, z], [x+\tfrac{1}{2}, y+\tfrac{1}{2}, z], [x, y+\tfrac{1}{2}, z+\tfrac{1}{2}],$ or $[x+\tfrac{1}{2}, y, z+\tfrac{1}{2}]$, and $[X, Y, Z]$ are coordinates referred to the unit cell (4), these being integers for all four choices of χ, η, ζ as can be seen from (5).

To complete the proof, we note that

$$(F/P)\begin{pmatrix} X \\ Y \\ Z \end{pmatrix} = \tfrac{1}{2}\begin{pmatrix} Y+Z \\ Z+Y \\ X+Y \end{pmatrix} = \begin{pmatrix} x \\ y \\ z \end{pmatrix} \text{ or } \begin{pmatrix} x+\tfrac{1}{2} \\ y+\tfrac{1}{2} \\ z \end{pmatrix} \text{ or } \begin{pmatrix} x \\ y+\tfrac{1}{2} \\ z+\tfrac{1}{2} \end{pmatrix} \text{ or } \begin{pmatrix} x+\tfrac{1}{2} \\ y \\ z+\tfrac{1}{2} \end{pmatrix}$$

for any triplet of integers X, Y, Z. The most important application of (4) is to the face-centred cubic lattice, in which case $\mathbf{A}, \mathbf{B}, \mathbf{C}$ define a primitive rhombohedral cell of angle $\cos^{-1}\tfrac{1}{2}$.

The f.c.c. (111) plane has the same arrangement of lattice points as that of the hexagonal basal plane (Fig. 6.15). This may be proved most easily by first constructing the primitive cubic (111) plane following the procedure of section 5.3, and then introducing all the face-centred lattice points which lie in this plane. For convenience we work with the plane of equation $x+y-z=0$ (symmetrically equivalent to $x+y+z=0$), which passes through the lattice points

$$[0, 0, 0], \quad [1, 0, 1], \quad [0, 1, 1]. \tag{6}$$

These define the two independent lattice vectors

$$\mathbf{B}=\mathbf{i}+\mathbf{k}, \quad \mathbf{A}=\mathbf{j}+\mathbf{k} \tag{7}$$

emanating from $[0, 0, 0]$, which are of equal length and inclined at an angle $\cos^{-1}\tfrac{1}{2}=60°$. Now the face-centred points $[\tfrac{1}{2}, 0, \tfrac{1}{2}]$, $[0, \tfrac{1}{2}, \tfrac{1}{2}]$ also lie in this plane, so defining two lattice vectors $\tfrac{1}{2}\mathbf{A}$, $\tfrac{1}{2}\mathbf{B}$ which generate all the lattice vectors, and therefore all the lattice points, in this plane. Since the generating cell is a rhombus of angle $60°$, it produces the hexagonal net shown in Fig. 6.15.

The next two parallel planes pass through $[0, 0, 1]$, $[0, 0, 2]$ respectively, and therefore they have the equations

$$x+y-z=-1 \text{ (plane 2)}, \quad x+y-z=-2 \text{ (plane 3)}$$

respectively. These planes possess the same arrangement of lattice points as that of the original plane (plane 1), and they project upon it as marked by the points ②, ③ in Fig. 6.17, as may be readily verified. Accordingly, the stacking pattern of the f.c.c. (111) planes is ...123123... compared with ...121212... for the c.p.h. (001) planes*. These alternative patterns are equally efficient for the close-packing of equal rigid spheres, i.e. each utilises a fraction 0.74 of space occupied, compared with 0.68 for the body-centred cubic packing, 0.52 for primitive cubic packing and 0.34 for diamond packing.

The diamond structure is generated by a face-centred cubic cell containing carbon (silicon, germanium) atoms at $[0, 0, 0]$, $[\tfrac{1}{4}, \tfrac{1}{4}, \tfrac{1}{4}]$ relative to each lattice point. This implies the well-known stereochemical picture, according to which each atom lies at the centroid of a tetrahedron defined by four

*Conventionally written (0001) planes (see section 7.4).

neighbouring atoms. Thus, the lattice point $[x, y, z]$ has three neighbouring lattice points $[x-\frac{1}{2}, y-\frac{1}{2}, z]$, $[x-\frac{1}{2}, y, z-\frac{1}{2}]$, $[x, y-\frac{1}{2}, z-\frac{1}{2}]$;   (8)
therefore there are atoms located at

$$[x, y, z], \quad [x+\tfrac{1}{4}, y+\tfrac{1}{4}, z+\tfrac{1}{4}] \tag{9}$$

and at

$$[x-\tfrac{1}{4}, y-\tfrac{1}{4}, z+\tfrac{1}{4}], \quad [x-\tfrac{1}{4}, \quad y+\tfrac{1}{4}, z-\tfrac{1}{4}], \quad [x+\tfrac{1}{4}, y-\tfrac{1}{4}, z-\tfrac{1}{4}]. \tag{10}$$

It will be readily verified that the first atom is equidistant from the other four, and that these four are equidistant from each other. Despite its stereographic simplicity, diamond has one of the most complex possible space groups involving both screw axes and 'diamond' glide planes (Chapter 13).

7.3 END-CENTRED CELLS

As will be seen later, a more valuable transformation in practice than either (1) or (4) is

$$\mathbf{A} = \tfrac{1}{2}(\mathbf{a}+\mathbf{b}), \quad \mathbf{B} = \tfrac{1}{2}(\mathbf{a}-\mathbf{b}), \quad \mathbf{C} = \mathbf{c} \tag{11}$$

which converts the face-centred lattice into the body-centred lattice (Fig. 7.1). Alternatively expressed, it induces the coordinate transformation

$$[x, y, z], \quad [x+\tfrac{1}{2}, y+\tfrac{1}{2}, z] \rightarrow [X, Y, Z],$$
$$[x+\tfrac{1}{2}, y, z+\tfrac{1}{2}], \quad [x, y+\tfrac{1}{2}, z+\tfrac{1}{2}] \rightarrow [X+\tfrac{1}{2}, Y+\tfrac{1}{2}, Z+\tfrac{1}{2}],$$

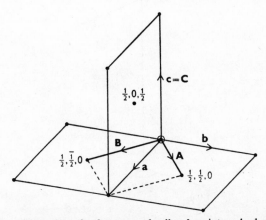

Fig. 7.1—Transformation of a face centred cell **a**, **b**, **c** into a body-centred cell **A**, **B**, **C**. The face centred lattice points $[\frac{1}{2}, \frac{1}{2}, 0]$; $[\frac{1}{2}, \frac{1}{2}, 0]$ become cornerpoints of the new cell, whereas the face-centred point $[\frac{1}{2}, 0, \frac{1}{2}]$ becomes a body-centred point of the new cell.

where X, Y, Z are integers, the appropriate transformation matrices being

$$(I/F) = \begin{pmatrix} 1 & 1 & 0 \\ 1 & \bar{1} & 0 \\ 0 & 0 & 1 \end{pmatrix}, \quad (F/I) = \tfrac{1}{2} \begin{pmatrix} 1 & 1 & 0 \\ 1 & \bar{1} & 0 \\ 0 & 0 & 1 \end{pmatrix}.$$

An end-centred lattice can be envisaged consisting of the arrays $[x, y, z]$ and $[x+\tfrac{1}{2}, y+\tfrac{1}{2}, z]$, but this could evidently be referred to a primitive cell by using (11). However, a doubly end-centred lattice, i.e. consisting of the three arrays $[x, y, z]$, $[x+\tfrac{1}{2}, y+\tfrac{1}{2}, z]$, and either $[x+\tfrac{1}{2}, y, z+\tfrac{1}{2}]$ or $[x, y+\tfrac{1}{2}, z+\tfrac{1}{2}]$ would not be admissible.

7.4 RHOMBOHEDRAL CELLS

There exists a non-primitive hexagonal cell which does not fall within any of the preceding types. It is perhaps best approached indirectly: start with a primitive rhombohedral cell defined by three symmetrically oriented vectors **A, B, C** of equal length, and note that the triangle joining $[1, 0, 0]$, $[0, 1, 0]$, $[0, 0, 1]$ is equilateral, thereby generating a hexagonal net of equation $X+Y+Z=1$; if so, parallel planes $X+Y+Z=0, 2, 3$ pass respectively through $[0, 0, 0]$, $[0, 0, 2]$, $[1, 1, 1]$. By symmetry, the planes $X+Y+Z=0, 3$ lie perpendicular to the axis [111], showing that the lattice can be referred to a hexagonal cell containing interior points trisecting the long diagonal (Fig. 7.2). This cell is constructed from the vectors

$$\mathbf{a} = \mathbf{A} - \mathbf{B}, \quad \mathbf{b} = \mathbf{B} - \mathbf{C}, \quad \mathbf{c} = \mathbf{A} + \mathbf{B} + \mathbf{C}, \tag{12}$$

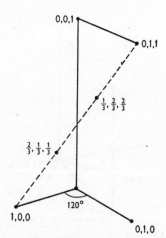

Fig. 7.2—Non-primitive hexagonal cell, i.e. R, showing location of interior lattice points. By contrast to the c.p.h. structure, the combined arrays x, y, z; $x+\tfrac{2}{3}, y+\tfrac{1}{3}, z+\tfrac{1}{3}$; $x+\tfrac{1}{3}, y+\tfrac{2}{3}, z+\tfrac{2}{3}$ form a space lattice.

where

$$\mathbf{c}.\mathbf{a} = \mathbf{c}.\mathbf{b} = 0, \quad \mathbf{a}.\mathbf{b} = ab \cos \frac{2\pi}{3}, \quad a = b,$$

and it contains interior points at

$$\mathbf{a} + \tfrac{1}{3}(\mathbf{b} + \mathbf{c} - \mathbf{a}), \quad \mathbf{a} + \tfrac{2}{3}(\mathbf{b} + \mathbf{c} - \mathbf{a}). \tag{13}$$

Conversely, given a non-primitive hexagonal cell containing the interior points (13), it generates a lattice which may be referred to the primitive rhombohedral cell

$$\mathbf{A} = \tfrac{1}{3}(2\mathbf{a} + \mathbf{b} + \mathbf{c}), \quad \mathbf{B} = \tfrac{1}{3}(-\mathbf{a} + \mathbf{b} + \mathbf{c}), \quad \mathbf{C} = \tfrac{1}{3}(-\mathbf{a} - 2\mathbf{b} + \mathbf{c}), \tag{14}$$

the appropriate transformation matrices being

$$(R/H) = \begin{pmatrix} 1 & 0 & 1 \\ \bar{1} & 1 & 1 \\ 0 & \bar{1} & 1 \end{pmatrix}, \quad (H/R) = \tfrac{1}{3} \begin{pmatrix} 2 & \bar{1} & \bar{1} \\ 1 & 1 & \bar{2} \\ 1 & 1 & 1 \end{pmatrix}.$$

A contrast to the rhombohedral space lattice is provided by the geometry of the close-packed hexagonal structure. As already noted, this contains two identical atoms at $[0, 0, 0]$, $[\tfrac{2}{3}, \tfrac{1}{3}, \tfrac{1}{2}]$ within the primitive hexagonal cell. It may now be queried whether or not the combined arrays

$$[x, y, z], \quad [x + \tfrac{2}{3}, y + \tfrac{1}{3}, z + \tfrac{1}{2}]; \quad x, y, z = 0, \ \pm 1, \ \pm 2, \ldots \tag{15}$$

together form a space lattice. Mathematically expressed, we look for three non-coplanar vectors $\mathbf{A}, \mathbf{B}, \mathbf{C}$ which generate (15) by the superposition process

$$X\mathbf{A} + Y\mathbf{B} + Z\mathbf{C}; \quad X, Y, Z = 0, \ \pm 1, \ \pm 2, \ldots \tag{16}$$

However, the reader will readily verify that such a triplet of vectors does not exist, and therefore our original model of the structure still stands.*

The hexagonal net (Fig. 6.3) is generated by vectors $\mathbf{a}, \mathbf{b}$ forming the sides of a rhombus of angle $120°$. There exists, however, a third vector $\mathbf{e}$ equivalent to these by a rotation through $120°$ about the 3-fold or 6-fold symmetry axis lying perpendicular to the net, and points along this direction satisfy the parametric equation $x = y = -t$. The line of equation $hx + ky = 1$ makes an

---

*The two-dimensional arrays $[x, y, 0]$, $[x + \tfrac{2}{3}, y + \tfrac{1}{3}, 0]$; $x, y = 0, \pm 1, \pm 2, \ldots$, define respectively the ①-points and ②-points of Fig. 6.16, which are seen to be generated by a rhombus of ①-points enclosing a ②-point. Clearly this cell could not be transformed into any alternative parallelogram containing no interior points on the lines of section 7.1. In fact, the combined arrays may be arranged into an hexagonal honeycomb pattern as shown in Fig. 6.19.

intercept $t = -(h+k)^{-1}$ along the **e**-direction, compared with intercepts h^{-1}, k^{-1} along the **a**, **b**-directions respectively. This means that the Miller indices (hkl) of a plane should preferably be written $(h, k, \overline{h+k}, l)$ to bring out crystallographic equivalence. Thus, for instance, the six planes

$$(100), (010), (\overline{1}10); \quad (\overline{1}00), (0\overline{1}0), (1\overline{1}0) \tag{17}$$

exhibit a regular hexagonal symmetry pattern made much more apparent by writing them as

$$(10\overline{1}0), (01\overline{1}0), (\overline{1}100); \quad (\overline{1}010), (0\overline{1}10), (1\overline{1}00). \tag{18}$$

According to convention, the six brackets (18) are all embraced by the form $\{10\overline{1}0\}$, meaning that they can all be generated by permuting the first three integers, 1, 0, $\overline{1}$. No such form would be available for (17). In line with this symbolism for planes, we write [uvwr] for directions, these indices being related to [pqr] by

$$u = \tfrac{1}{3}(2p - q), \quad v = \tfrac{1}{3}(2q - p), \quad w = -(u + v) = -\tfrac{1}{3}(p + q), \quad r = r$$
$$p = u - w, \quad q = v - w, \quad r = r. \tag{19}$$

It is sometimes necessary to transform from the hexagonal indices $(hk.l)$ to rhombohedral indices (HKL) and vice versa: use the relation $[x, y, z] = (H/R)$ $[X, Y, Z]$ to eliminate x, y, z in favour of X, Y, Z from the expression $hx + ky + lz$, and so transform it into $HX + KY + LZ$ by collecting terms, whence

$$H = h - k, \quad h = \tfrac{1}{3}(2H + K + L),$$
$$K = k - l, \quad k = \tfrac{1}{3}(-H + K + L), \tag{20}$$
$$L = h + k + l, l = \tfrac{1}{3}(-H - 2K + L).$$

7.5 SYMMETRY OF NON-PRIMITIVE BRAVAIS CELLS

We have noted that the primitive cubic lattice has a holohedral symmetry $\frac{4}{m}\,\overline{3}\,\frac{2}{m}$ with respect to any lattice point as fixed point. This also holds for the body-centred cubic lattice, because the eight body-centred points $[\pm\tfrac{1}{2}, \pm\tfrac{1}{2}, \pm\tfrac{1}{2}]$ transform into each other under the operations of $\left\{\frac{4}{m}\,\overline{3}\,\frac{2}{m}\right\}$. Similarly, the twelve face-centred points $[\pm\tfrac{1}{2}, \pm\tfrac{1}{2}, 0]$, $[0, \pm\tfrac{1}{2}, \pm\tfrac{1}{2}]$, $[\pm\tfrac{1}{2}, 0, \pm2]$ transform into each other under the operations of $\left\{\frac{4}{m}\,\overline{3}\,\frac{2}{m}\right\}$. No other independent possibilities exist. Thus, given a crystal characterised macroscopically by any cubic symmetry, we can infer that its underlying space lattice must be one of the three in question, but no further information could be deduced from macroscropic evidence alone. Both the body-centred cubic and face-centred cubic lattices can be referred to a primitive rhombohedral cell, as previously mentioned, but the advantages of a non-primitive cell exhibiting cubic symmetry far outweigh those of a primitive cell exhibiting only rhombohedral symmetry. More precisely, it would not be at all apparent from these rhombohedral cells that the generated space lattices exhibit cubic symmetry. Distorting one edge of a body-centred cubic cell produces a body-centred tetragonal cell compatible with all the tetragonal symmetries. The face-centred tetragonal cell can be referred to a body-centred tetragonal cell by virtue of (11). A body-centred orthorhombic cell exists. But, unlike the tetragonal cell, face-centred and end-centred orthorhombic cells also exist since they cannot be transformed away without reducing the cell symmetry. A body-centred monoclinic cell exists, but nothing is gained by centring any of the faces. The non-primitive hexagonal cell, i.e. the rhombohedral cell, has already been considered. So we arrive at the results displayed in Table 9.1.

Chapter 8

Translation Groups

8.1 TRANSLATION OPERATORS

It is convenient to think of lattice vectors as produced by the action of translation operators, which play an analogous mathematical role to that of the point-group operators. Thus, we introduce operators $\mathscr{A}$, $\mathscr{B}$, $\mathscr{C}$ with the properties

$$\mathscr{A}\mathbf{R}=\mathbf{R}+\mathbf{a}, \quad \mathscr{B}\mathbf{R}=\mathbf{R}+\mathbf{b}, \quad \mathscr{C}\mathbf{R}=\mathbf{R}+\mathbf{c}, \tag{1}$$

where $\mathbf{R}$ is an arbitrary vector and $\mathbf{a}$, $\mathbf{b}$, $\mathbf{c}$ are the generating vectors of a space lattice. It follows from (1) that

$$\mathscr{A}^\lambda \mathscr{B}^\mu \mathscr{C}^\nu \mathbf{R}=\mathbf{R}+\lambda\mathbf{a}+\mu\mathbf{b}+\nu\mathbf{c}; \quad \lambda, \mu, \nu = 0, \pm 1, \pm 2, \ldots \tag{2}$$

and from (2) that

$$\mathscr{A}^\lambda \mathscr{B}^\mu \mathscr{C}^\nu = \mathscr{B}^\mu \mathscr{A}^\lambda \mathscr{C}^\nu, \text{ etc.} \tag{3}$$

These properties imply that all the powers and products of $\mathscr{A}$, $\mathscr{B}$, $\mathscr{C}$ form an infinite Abelian group, i.e. the primitive translation group symbolised by

$$\{\mathscr{T}\}=\{I, \mathscr{A}, \mathscr{B}, \mathscr{C}, \ldots \mathscr{A}^\lambda \mathscr{B}^\mu \mathscr{C}^\nu, \ldots\}. \tag{4}$$

Our symbolism indicates that $\mathscr{T}$ stands for any operator of the form (3), and this produces a lattice vector $\mathbf{t}$ in accordance with

$$\mathscr{T}\mathbf{R}=\mathbf{R}+\mathbf{t}. \tag{5}$$

Body-centred translation operators have the form $\mathscr{A}^{\frac{1}{2}}\mathscr{B}^{\frac{1}{2}}\mathscr{C}^{\frac{1}{2}}\mathscr{T}$, and hence they are all included in the infinite set

$\mathscr{A}^{1/2}\mathscr{B}^{1/2}\mathscr{C}^{1/2}\{\mathscr{T}\}$. The combined sets

$$\{\mathscr{T}\}, \quad \mathscr{A}^{1/2}\mathscr{B}^{1/2}\mathscr{C}^{1/2}\{\mathscr{T}\} \tag{6}$$

constitute a group, as follows essentially from the closure properties:

$$\mathscr{A}^{1/2}\mathscr{B}^{1/2}\mathscr{C}^{1/2} \cdot \mathscr{A}^{1/2}\mathscr{B}^{1/2}\mathscr{C}^{1/2} = \mathscr{A}\mathscr{B}\mathscr{C} \subset \{\mathscr{T}\}$$

$$(\mathscr{A}^{1/2}\mathscr{B}^{1/2}\mathscr{C}^{1/2})^{-1} = \mathscr{C}^{-1/2}\mathscr{B}^{-1/2}\mathscr{A}^{-1/2} = \mathscr{A}^{1/2}\mathscr{B}^{1/2}\mathscr{C}^{1/2} \cdot \mathscr{A}^{-1}\mathscr{B}^{-1}\mathscr{C}^{-1}$$

$$\subset \mathscr{A}^{1/2}\mathscr{B}^{1/2}\mathscr{C}^{1/2}\{\mathscr{T}\}.$$

This is the body-centred translation group, more compactly written

$$\{\mathscr{I}\} = \{I, \mathscr{A}^{1/2}\mathscr{B}^{1/2}\mathscr{C}^{1/2}\}\{\mathscr{T}\}. \tag{7}$$

Similarly, we may construct the end-centred translation group

$$\{\mathscr{E}\} = \{I, \mathscr{A}^{1/2}\mathscr{B}^{1/2}\}\{\mathscr{T}\} \tag{8}$$

and the face-centred translation group

$$\{\mathscr{F}\} = \{I, \mathscr{A}^{1/2}\mathscr{B}^{1/2}, \mathscr{B}^{1/2}\mathscr{C}^{1/2}, \mathscr{C}^{1/2}\mathscr{A}^{1/2}\}\{\mathscr{T}\} \tag{9}$$

on bearing in mind the closure properties

$$\mathscr{A}^{1/2}\mathscr{B}^{1/2} \cdot \mathscr{A}^{1/2}\mathscr{B}^{1/2} = \mathscr{A}\mathscr{B} \subset \{\mathscr{T}\},$$

$$\mathscr{A}^{1/2}\mathscr{B}^{1/2} \cdot \mathscr{B}^{1/2}\mathscr{C}^{1/2} = \mathscr{A}^{1/2}\mathscr{B}\mathscr{C}^{1/2} = \mathscr{C}^{1/2}\mathscr{A}^{1/2}\mathscr{B} \subset \mathscr{C}^{1/2}\mathscr{A}^{1/2}\{\mathscr{T}\},$$

etc.

Finally, there exists a doubly-centred translation group

$$\{\mathscr{R}\} \equiv \{I, \mathscr{A}^{2/3}\mathscr{B}^{1/3}\mathscr{C}^{1/3}, \mathscr{A}^{1/3}\mathscr{B}^{2/3}\mathscr{C}^{2/3}\}\{\mathscr{T}\}, \tag{10}$$

as may be verified from the closure properties

$$(\mathscr{A}^{2/3}\mathscr{B}^{1/3}\mathscr{C}^{1/3})^2 = \mathscr{A}^{4/3}\mathscr{B}^{2/3}\mathscr{C}^{2/3} = \mathscr{A}^{1/3}\mathscr{B}^{2/3}\mathscr{C}^{2/3} \cdot \mathscr{A}$$

$$\subset \mathscr{A}^{1/3}\mathscr{B}^{2/3}\mathscr{C}^{2/3}\{\mathscr{T}\},$$

$$(\mathscr{A}^{2/3}\mathscr{B}^{1/3}\mathscr{C}^{1/3})^{-1} = \mathscr{C}^{-1/3}\mathscr{B}^{-1/3}\mathscr{A}^{-2/3} = \mathscr{A}^{1/3}\mathscr{B}^{2/3}\mathscr{C}^{2/3} \cdot \mathscr{A}^{-1}\mathscr{B}^{-1}\mathscr{C}^{-1}$$

$$\subset \mathscr{A}^{1/3}\mathscr{B}^{2/3}\mathscr{C}^{2/3}\{\mathscr{T}\}, \text{ etc.,}$$

having obvious reference to the rhombohedral (doubly-centred hexagonal) space lattice.

8.2 COUPLING BETWEEN $\{\mathscr{T}\}$ AND $\{G\}$

It has already been noted that every Bravais lattice has a point-group symmetry with respect to any of its lattice points 0. This implies that the relevant translation group $\{\mathscr{T}\}$ must be compatible with the relevant point-group $\{G\}$. To arrive at a suitable coupling condition, we note that $Gt_1(=t_2)$ is a lattice vector emanating from 0 if t_1 is a lattice vector emanating from 0, where $|t_1| = |t_2|$. This property may be expressed in the convenient operational form

$$G_i\mathscr{T}_1\,G_i^{-1}=\mathscr{T}_2; \quad \mathscr{T}_1, \mathscr{T}_2 \subset \{\mathscr{T}\}, \quad G_i \subset \{\mathscr{G}\}. \tag{11}$$

as follows from

$$G_i\mathscr{T}_1 G_i^{-1}\mathbf{R} = G_i\mathscr{T}_1.G_i^{-1}\mathbf{R} = G_i.\mathscr{T}_1\,(G_i^{-1}\mathbf{R})= G_i(G_i^{-1}\mathbf{R}+t_1)$$

$$=\mathbf{R}+G_it_1=\mathbf{R}+t_2=\mathscr{T}_2\mathbf{R}.$$

Relation (11) provides a mathematical coupling condition between $\{G\}$ and $\{\mathscr{T}\}$ which serves as the foundation for space-group theory.

No difficulty arises in extending (11) to centred space lattices. Body-centred cubic, tetragonal, orthorhombic and monoclinice lattices obey the supplementary condition

$$G_i\mathscr{A}^{1/2}\mathscr{B}^{1/2}\mathscr{C}^{1/2}G_i^{-1}=\mathscr{A}^{\pm 1/2}\mathscr{B}^{\pm 1/2}\mathscr{C}^{\pm 1/2} \subset \mathscr{A}^{1/2}\mathscr{B}^{1/2}\mathscr{C}^{1/2}\{\mathscr{T}\} \subset \{\mathscr{I}\}. \tag{12}$$

Face-centred cubic and orthorhombic lattices obey the supplementary condition

$$G_i\mathscr{A}^{1/2}\mathscr{B}^{1/2}G_i^{-1}=\mathscr{A}^{\pm 1/2}\mathscr{B}^{\pm 1/2} \text{ or } \mathscr{B}^{\pm 1/2}\mathscr{C}^{\pm 1/2} \text{ or } \mathscr{C}^{\pm 1/2}\mathscr{A}^{\pm 1/2}$$

$$\subset \{\mathscr{A}^{1/2}\mathscr{B}^{1/2}, \mathscr{B}^{1/2}\mathscr{C}^{1/2}, \mathscr{C}^{1/2}\mathscr{A}^{1/2}\}\{\mathscr{T}\} \subset \{\mathscr{F}\}. \tag{13}$$

End-centred lattices obey the supplementary condition

$$G_i\mathscr{A}^{1/2}\mathscr{B}^{1/2}G_i^{-1}=\mathscr{A}^{\pm 1/2}\mathscr{B}^{\pm 1/2} \subset \mathscr{A}^{1/2}\mathscr{B}^{1/2}\{\mathscr{T}\} \subset \{\mathscr{E}\}. \tag{14}$$

Finally, doubly-centred hexagonal lattices obey the supplementary condition

$$G_i\mathscr{A}^{2/3}\mathscr{B}^{1/3}\mathscr{C}^{1/3}G_i^{-1}=\{\mathscr{A}^{2/3}\mathscr{B}^{1/3}\mathscr{C}^{1/3}, \mathscr{A}^{1/3}\mathscr{B}^{2/3}\mathscr{C}^{2/3}\}\{\mathscr{T}\} \subset \{\mathscr{R}\}. \tag{15}$$

Unless otherwise stated, $\{G\}$ refers to the holohedral point-group symmetry of the space lattice. It could, however, be any lower symmetry appertaining to the crystal system in question as displayed in Table 9.1.

8.3 COLOUR TRANSLATION OPERATORS

Corresponding with $\mathscr{A}, \mathscr{B}, \mathscr{C}$ we introduce the colour translation operators $S\mathscr{A}, S\mathscr{B}, S\mathscr{C}$ where $S\mathscr{A} =\mathscr{A}S$, etc. Consequently

$$S\mathscr{A}.S\mathscr{B} = S^2\mathscr{A}\mathscr{B} =\mathscr{A}\mathscr{B},$$

and more generally

$$(S\mathcal{A})^{\lambda}(S\mathcal{B})^{\mu}(S\mathcal{C}^{\nu}) = \mathcal{A}^{\lambda}\mathcal{B}^{\mu}\mathcal{C}^{\nu} \quad ; \quad \lambda + \mu + \nu = 2n$$
$$= S\mathcal{A}^{\lambda}\mathcal{B}^{\mu}\mathcal{C}^{\nu}; \quad \lambda + \mu + \nu = 2n+1 \tag{16}$$

on bearing in mind $S^{2n} = I$, $S^{2n+1} = S$. Clearly the infinite set of operators

$$\{S^{\lambda+\mu+\nu}\mathcal{A}^{\lambda}\mathcal{B}^{\mu}\mathcal{C}^{\nu}\}; \quad \lambda, \mu, \nu = 0, \pm 1, \pm 2, \ldots \tag{17}$$

forms an Abelian group, with the property

$$\Leftrightarrow \{\mathcal{A}^{\lambda}\mathcal{B}^{\mu}\mathcal{C}^{\nu}\}; \qquad \lambda, \mu, \nu = 0, \pm 1, \pm 2, \ldots$$

This property also holds for the set

$$\{S^{\mu+\nu}\mathcal{A}^{\lambda}\mathcal{B}^{\mu}\mathcal{C}^{\nu}\}; \qquad \lambda, \mu, \nu = 0, \pm 1, \pm 2, \ldots \tag{18}$$

generated from the triplet $\mathcal{A}, S\mathcal{B}, S\mathcal{C}$; and for the set

$$\{S^{\nu}\mathcal{A}^{\lambda}\mathcal{B}^{\mu}\mathcal{C}^{\nu}\}; \qquad \lambda, \mu, \nu = 0, \pm 1, \pm 2, \ldots \tag{19}$$

generated from the triplet $\mathcal{A}, \mathcal{B}, S\mathcal{C}$. These are the colour translation groups having no particular symmetry features.

The operator $S\mathcal{A}^{\lambda}\mathcal{B}^{\mu}\mathcal{C}^{\nu}$ acts upon an arbitrary lattice vector $\mathbf{R}$ to produce

$$S\mathcal{A}^{\lambda}\mathcal{B}^{\mu}\mathcal{C}^{\nu}\mathbf{R} = S(\mathbf{R} + \lambda\mathbf{a} + \mu\mathbf{b} + \nu\mathbf{c}), \tag{20}$$

which signifies a colour switch at the terminus of $\mathbf{R} + \lambda\mathbf{a} + \mu\mathbf{b} + \nu\mathbf{c}$ compared with the terminus of $\mathbf{R}$. This interpretation readily enables us to construct colour space lattices produced by the sets (17)–(19). Two-dimensional colour lattices produced by the groups

$$\{S^{\lambda+\mu}\mathcal{A}^{\lambda}\mathcal{B}^{\mu}\}, \qquad \{S^{\mu}\mathcal{A}^{\lambda}\mathcal{B}^{\mu}\}; \qquad \lambda, \mu = 0, \pm 1, \pm 2, \ldots \tag{21}$$

appear in Fig. 8.1, 8.2 respectively. At first sight these two lattices seem to be distinct, but closer inspection shows that they are essentially equivalent. More precisely, if we transform the generating lattice vectors from $\mathbf{a}, \mathbf{b}$ to $\bar{\mathbf{a}} = \mathbf{a} + \mathbf{b}, \bar{\mathbf{b}} = \mathbf{b}$ in Fig. 8.1, then $\bar{\mathbf{a}}, \bar{\mathbf{b}}$ play an analogous colour role to that of $\mathbf{a}, \mathbf{b}$ in Fig. 8.2. Alternatively, the transformation

$$\bar{\mathcal{A}} = \mathcal{A}\mathcal{B}, \bar{\mathcal{B}} = \mathcal{B} \text{ i.e. } \mathcal{A} = \bar{\mathcal{A}}\bar{\mathcal{B}}^{-1}, \mathcal{B} = \bar{\mathcal{B}},$$

yields

$$S^{\lambda+\mu}\mathcal{A}^{\lambda}\mathcal{B}^{\mu} = S^{\lambda+\mu}\bar{\mathcal{A}}^{\lambda}\bar{\mathcal{B}}^{\mu-\lambda} = S^{2\lambda}.S^{\mu-\lambda}\bar{\mathcal{A}}^{\lambda}\bar{\mathcal{B}}^{\mu-\lambda} = S^{\mu-\lambda}\bar{\mathcal{A}}^{\lambda}\bar{\mathcal{B}}^{\mu-\lambda},$$

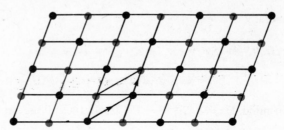

Fig. 8.1—Two-dimensional colour lattice produced by $S\mathscr{A}$, $S\mathscr{B}$ showing transformation from **a**, **b** to **a**, **b** (the arrowed vectors respectively).

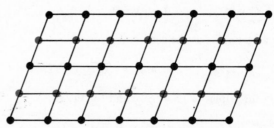

Fig. 8.2—Two-dimensional colour lattice produced by $S\mathscr{A}$, $\mathscr{B}$ with generating vectors **a**, **b** equivalent to **a**, **b** of Fig. 8.1.

showing that

$$\{S^{\lambda+\mu}\mathscr{A}^{\lambda}\mathscr{B}^{\mu}\} \Leftrightarrow \{S^{\mu}\mathscr{A}^{\lambda}\mathscr{B}^{\mu}\}. \tag{22}$$

It follows that

$$S^{\lambda+\mu+\nu}\mathscr{A}^{\lambda}\mathscr{B}^{\mu}\mathscr{C}^{\nu}\} \Leftrightarrow \{S^{\mu+\nu}\mathscr{A}^{\lambda}\mathscr{B}^{\mu}\mathscr{C}^{\nu} \Leftrightarrow \{S^{\nu}\mathscr{A}^{\lambda}\mathscr{B}^{\mu}\mathscr{C}^{\nu}\}, \tag{24}$$

which signifies the existence of only one independent colour triclinic space lattice.

It is intuitively clear that colour lattice points transform into colour lattice points under the relevant point-group operations.

To express this in mathematical form, we first note that

$$G_i S = S G_i, \text{ i.e. } G_i S G_i^{-1} = S \tag{25}$$

from which there follows

$$G_i S \mathscr{T}_1 G_i^{-1} = G_i S G_i^{-1} . G_i \mathscr{T}_1 G_i^{-1} = S \mathscr{T}_2 \tag{26}$$

by virtue of (11) above. Condition (26) holds automatically for the colour translation groups listed in Table 8.1, provided that $\{G\}$ conforms to the appropriate crystal system or systems indicated.

Table 8.1—Colour Unit Calls

Lattice type	Crystal system	Colour translation group
P_s	triclinic	$(I, S\mathscr{C}^{1/2})\{\mathscr{T}\}$
P_c	monoclinic orthorhombic tetragonal hexagonal	$(I, S\mathscr{C}^{1/2})\{\mathscr{T}\}$
P_a	monoclinic	$(I, S\mathscr{A}^{1/2})\{\mathscr{T}\}$
P_I	monoclinic orthorhombic tetragonal cubic	$(I, S\mathscr{A}^{1/2}\mathscr{B}^{1/2}\mathscr{C}^{1/2})\{\mathscr{T}\}$
P_C	orthorhombic tetragonal	$(I, S\mathscr{A}^{1/2}\mathscr{B}^{1/2})\{\mathscr{T}\}$
I_c	monoclinic orthorhombic tetragonal	$\{(I, \mathscr{A}^{1/2}\mathscr{B}^{1/2}\mathscr{C}^{1/2}) + S\mathscr{C}^{1/2}(I, \mathscr{A}^{1/2}\mathscr{B}^{1/2}\mathscr{C}^{1/2})\}\{\mathscr{T}\}$
I_a	monoclinic	$\{(I, \mathscr{A}^{1/2}\mathscr{B}^{1/2}\mathscr{C}^{1/2}) + S\mathscr{A}^{1/2}(I, \mathscr{A}^{1/2}\mathscr{B}^{1/2}\mathscr{C}^{1/2})\}\{\mathscr{T}\}$
C_A	orthorhombic	$\{(I, \mathscr{A}^{1/2}\mathscr{B}^{1/2}) + S\mathscr{B}^{1/2}\mathscr{C}^{1/2}(I, \mathscr{A}^{1/2}\mathscr{B}^{1/2})\}\{\mathscr{T}\}$
C_c	orthorhombic	$\{(I, \mathscr{A}^{1/2}\mathscr{B}^{1/2}) + S\mathscr{C}^{1/2}(I, \mathscr{A}^{1/2}\mathscr{B}^{1/2})\}\{\mathscr{T}\}$
C_a	orthorhombic	$\{(I, \mathscr{A}^{1/2}\mathscr{B}^{1/2}) + S\mathscr{A}^{1/2}(I, \mathscr{A}^{1/2}\mathscr{B}^{1/2})\}\{\mathscr{T}\}$
F_c	orthorhombic cubic	$\{(I, \mathscr{A}^{1/2}\mathscr{B}^{1/2}, \mathscr{B}^{1/2}\mathscr{C}^{1/2}, \mathscr{C}^{1/2}\mathscr{A}^{1/2}) + S\mathscr{C}^{1/2}(I, \mathscr{A}^{1/2}\mathscr{B}^{1/2}, \mathscr{B}^{1/2}\mathscr{C}^{1/2}, \mathscr{C}^{1/2}\mathscr{A}^{1/2})\}\{\mathscr{T}\}$
R_I	hexagonal	$\{(I, \mathscr{A}^{2/3}\mathscr{B}^{1/3}\mathscr{C}^{1/3}, \mathscr{A}^{1/3}\mathscr{B}^{2/3}\mathscr{C}^{2/3}) + S\mathscr{A}^{1/2}\mathscr{B}^{1/2}\mathscr{C}^{1/2}(I, \mathscr{A}^{2/3}\mathscr{B}^{1/3}\mathscr{C}^{1/3}, \mathscr{A}^{1/3}\mathscr{B}^{2/3}\mathscr{C}^{2/3})\}\{\mathscr{T}\}$

8.4 EDGE-CENTRING

A systematic method for the construction of coloured space lattices is by means of edge-centring. First, we note that the infinite translation group

$$\{I, \mathscr{C}^{1/2}\}\{\mathscr{T}\} \tag{27}$$

produces a generating unit cell defined by $\mathbf{a}, \mathbf{b}, \frac{1}{2}\mathbf{c}$. This preserves the symmetry of monoclinic, orthorhombic and tetragonal cells characterised by an axis $\mathbf{c}$ perpendicular to $\mathbf{a}, \mathbf{b}$. Accordingly the coloured translation group

$$\{I, S\mathscr{C}^{1/2}\}\{\mathscr{T}\} \tag{28}$$

produces these cells coloured along the principal axis. Such cells are symbolised P_c. For monoclinic symmetry there exists a second independent cell P_a, i.e. coloured along a secondary axis by

$$\{I, S\mathscr{A}^{1/2}\}\{\mathscr{T}\}, \tag{29}$$

which would be equivalent to P_c for orthorhombic symmetry and would not be admissible for tetragonal symmetry. The coloured triclinic cell is symbolised P_s, to signify the absence of a principal axis.

Two secondary cell edges are coloured by

$$\{I, S\mathscr{A}^{1/2}, S\mathscr{B}^{1/2}, \mathscr{A}^{1/2}\mathscr{B}^{1/2}\}\{\mathscr{T}\}, \tag{30}$$

i.e.

$$\{(I, \mathscr{A}^{1/2}\mathscr{B}^{1/2}) + S\mathscr{A}^{1/2}(I, \mathscr{A}^{1/2}\mathscr{B}^{1/2})\}\{\mathscr{T}\} \tag{30a}$$

producing a cell C_a. This reduces to P_s for triclinic symmetry and to P_a for monoclinic symmetry. Also, it is equivalent to P_c (see below) for tetragonal symmetry, by a transformation from edge-centred to end-centred lattice points, leaving orthorhombic symmetry as the only independent C_a symmetry. Three cell edges are coloured by

$$\{I, S\mathscr{A}^{1/2}, S\mathscr{B}^{1/2}, S\mathscr{C}^{1/2}, \mathscr{A}^{1/2}\mathscr{B}^{1/2}, \mathscr{B}^{1/2}\mathscr{C}^{1/2}, \mathscr{C}^{1/2}\mathscr{A}^{1/2}, S\mathscr{A}^{1/2}\mathscr{B}^{1/2}\mathscr{C}^{1/2}\}\{\mathscr{T}\}, \tag{31}$$

i.e.

$$\{(I, \mathscr{A}^{1/2}\mathscr{B}^{1/2}, \mathscr{B}^{1/2}\mathscr{C}^{1/2}, \mathscr{C}^{1/2}\mathscr{A}^{1/2}) + S\mathscr{C}^{1/2}(I, \mathscr{A}^{1/2}\mathscr{B}^{1/2}, \mathscr{B}^{1/2}\mathscr{C}^{1/2}, \mathscr{C}^{1/2}\mathscr{A}^{1/2})\}\{\mathscr{T}\}, \tag{31a}$$

producing a cell F_c. This provides new orthorhombic and cubic symmetries, but it is equivalent to I_c (see below) for monoclinic and tetragonal symmetries. Body-centred lattice points are coloured by

$$\{I, S\mathscr{A}^{1/2}\mathscr{B}^{1/2}\mathscr{C}^{1/2}\}\{\mathscr{T}\}, \tag{32}$$

producing a cell P_I. This provides new monoclinic, orthorhombic, tetragonal and cubic symmetries. Also, end-centred lattice points are coloured by

$$\{I, S\mathscr{A}^{1/2}\mathscr{B}^{1/2}\}\{\mathscr{T}\}, \tag{33}$$

producing a cell P_C. This provides new orthorhombic and tetragonal symmetries (i.e. on eliminating the tetragonal symmetry C_a above), but it reduces to P_a for monoclinic symmetry.

Body-centred lattices may be coloured along an edge by

$$\{(I, \mathscr{A}^{1/2}\mathscr{B}^{1/2}\mathscr{C}^{1/2}) + S\mathscr{C}^{1/2}(I, \mathscr{A}^{1/2}\mathscr{B}^{1/2}\mathscr{C}^{1/2})\}\{\mathscr{T}\}, \tag{34}$$

producing a cell I_c. This provides new monoclinic, orthorhombic, and tetragonal symmetries, provided we eliminate the monoclinic and tetragonal symmetries F_c above. However, there exists an additional monoclinic cell I_a, produced by

$$\{(I, \mathscr{A}^{1/2}\mathscr{B}^{1/2}\mathscr{C}^{1/2}) + S\mathscr{A}^{1/2}(I, \mathscr{A}^{1/2}\mathscr{B}^{1/2}\mathscr{C}^{1/2})\}\{\mathscr{T}\}, \tag{35}$$

which bears the same relation to I_c as does P_a to P_c above. End-centred lattices may be coloured along an edge by

$$\{(I, \mathscr{A}^{1/2}\mathscr{B}^{1/2}) + S\mathscr{C}^{1/2}(I, \mathscr{A}^{1/2}\mathscr{B}^{1/2})\}\{\mathscr{T}\}, \tag{36}$$

producing a cell C_c which partners C_a above for orthorhombic symmetry.

Two orthorhombic faces are coloured by

$$\{I, S\mathscr{B}^{1/2}\mathscr{C}^{1/2}, S\mathscr{C}^{1/2}\mathscr{A}^{1/2}, \mathscr{A}^{1/2}\mathscr{B}^{1/2}\}\{\mathscr{T}\}, \tag{37}$$

i.e.

$$\{(I, \mathscr{A}^{1/2}\mathscr{B}^{1/2}) + S\mathscr{B}^{1/2}\mathscr{C}^{1/2}(I, \mathscr{A}^{1/2}\mathscr{B}^{1/2})\}\{\mathscr{T}\}, \tag{38}$$

producing a cell C_A, and it will be noted that the more symmetric set

$$\{I, S\mathscr{B}^{1/2}\mathscr{C}^{1/2}, S\mathscr{C}^{1/2}\mathscr{A}^{1/2}, S\mathscr{A}^{1/2}\mathscr{B}^{1/2}\}\{\mathscr{T}\}, \tag{39}$$

does not constitute a group.

Finally, the group (31) also colours all three rhombohedral cell edges, producing a cell R_I. In hexagonal co-ordinates, (31) appears as

$$\{(I, \mathscr{A}^{2/3}\mathscr{B}^{1/3}\mathscr{C}^{1/3}, \mathscr{A}^{1/3}\mathscr{B}^{2/3}\mathscr{C}^{2/3}) + S\mathscr{A}^{1/2}\mathscr{B}^{1/2}\mathscr{C}^{1/3}(I, \mathscr{A}^{2/3}\mathscr{B}^{1/3}\mathscr{C}^{1/3},$$

$$\mathscr{A}^{1/3}\mathscr{B}^{2/3}\mathscr{C}^{2/3})\}\{\mathscr{T}\} \tag{40}$$

following (10) above.

Our conclusions are summarised in Table 8.1 and the accompanying unit cells are delineated in Fig. 8.3.

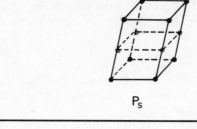

P_s

Triclinic

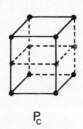
P_c

P_a

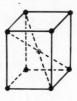
P_I

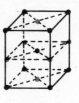
I_c

Monoclinic

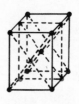
I_a

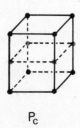
P_c

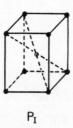
P_I

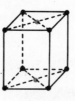
P_c

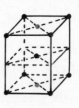
I_c

Orthorhombic

C_A

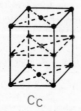
C_C

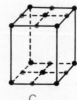
C_a

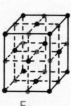
F_c

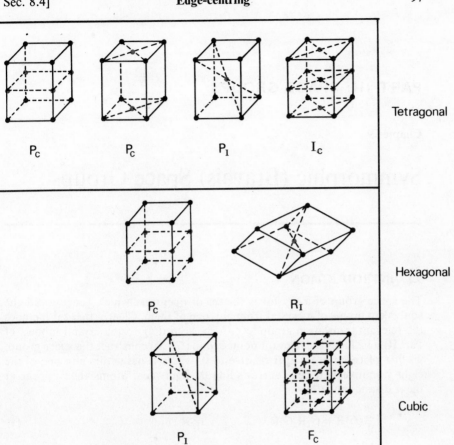

Fig. 8.3—Unit cells of colour Bravais lattices.

PART III: SPACE GROUPS

Chapter 9

Symmorphic (Bravais) Space Groups

9.1 INTRODUCTION

The space group of a crystal is the set of operators which generate all the equivalent atoms of a crystal from any one of them. Clearly this set includes the relevant translation group $\{\mathscr{T}\}$. Also, according to the crystal models of Part II, it includes the relevant point-group $\{G\}$. To construct the space group, we first place an atom at **R** relative to 0, where **R** lies within any one of the eight primitive unit cells surrounding 0. Equivalent atoms then appear at the n distinct positions

$$\{G\}\mathbf{R} = G_1\mathbf{R}, G_2\mathbf{R}, \ldots\ldots, \quad G_n\mathbf{R} \tag{1}$$

all of which lie on a sphere of radius R centred about 0. This set of atoms, by definition, constitutes the elementary motif pattern associated with 0. Operating upon the positions (1) by a translation operator $\mathscr{T}$, we generate further equivalent atoms at

$$\mathscr{T}\{G\}\mathbf{R} = \mathscr{T}G_1\mathbf{R}, \quad \mathscr{T}G_2\mathbf{R}, \ldots\ldots, \quad \mathscr{T}G_n\mathbf{R}. \tag{2}$$

i.e. at

$$\{G\}\mathbf{R} + \mathbf{t} = G_1\mathbf{R} + \mathbf{t}, \quad G_2\mathbf{R} + \mathbf{t}, \ldots\ldots, \quad G_n\mathbf{R} + \mathbf{t}, \tag{3}$$

thereby producing a replica of the original motif pattern translated through the lattice vector **t**. So proceeding for every choice $\mathscr{T}$ as given by (4), Chapter 8, we generate all the equivalent motif patterns—and hence all the equivalent atoms of the crystal—by the infinite set of operators

$$\{\mathscr{T}\} \otimes \{G\} = \{G\}, \mathscr{A}\{G\}, \mathscr{B}\{G\}, \mathscr{C}\{G\}, \ldots\ldots, \quad \mathscr{A}^\lambda\mathscr{B}^\upsilon\mathscr{C}^\varkappa\{G\}, \ldots \tag{4}$$

This is the symmorphic (Bravais) space group of the crystal, constructed so as to emphasise its translational symmetry. However, it fails to expose the crystal's rotational symmetry as a whole with respect to 0.

An alternative procedure is to place an atom at **R**, as before, and then generate equivalent atoms at

$$\{\mathcal{T}\}\mathbf{R} = \mathbf{R}, \mathscr{A}\mathbf{R}, \mathscr{B}\mathbf{R}, \mathscr{C}\mathbf{R}, \ldots, \quad \mathscr{A}^\lambda \mathscr{B}^\mu \mathscr{C}^\nu \mathbf{R}, \ldots \tag{5}$$

i.e. at

$$\mathbf{R}, \mathbf{R}+\mathbf{a}, \mathbf{R}+\mathbf{b}, \mathbf{R}+\mathbf{c}, \ldots, \quad \mathbf{R}+\lambda\mathbf{a}+\mu\mathbf{b}+\nu\mathbf{c}, \ldots \tag{6}$$

i.e. at positions **R** relative to each lattice point. Operating upon these positions by a point-group operator G_i which keeps 0 fixed, we generate further equivalent atoms at

$$G_i\{\mathcal{T}\}\mathbf{R} = G_i\mathbf{R}, G_i\mathscr{A}\mathbf{R}, \ldots \tag{7}$$

So proceeding for every choice of G_i as appears in (1), we generate all the equivalent atoms by the infinite set of operators

$$\{G\}\otimes\{\mathcal{T}\} = G_1\{\mathcal{T}\}, \quad G_2\{\mathcal{T}\}, \ldots\ldots, \quad G_n\{\mathcal{T}\}. \tag{8}$$

This is the symmorphic (Bravais) space group of the crystal, constructed so as to emphasise its rotational symmetry as a whole with respect to 0. However, it fails to expose the existence of a recurring motif pattern centred about each lattice point. There is no difficulty, however, in proving that (4), (8), comprise the same set of operators taken in different orders.

The set (8) forms a group by virtue of the properties:

1. $I \subset \{\mathcal{T}\}, \quad I \subset \{G\} \quad$ i.e. $\quad I \subset \{G\}\otimes\{\mathcal{T}\},$
2. $G_1\mathcal{T}_1 . G_2\mathcal{T}_2 = G_1 . \mathcal{T}_1 G_2 . \mathcal{T}_2 = G_1 . G_2\mathcal{T}_1' . \mathcal{T}_2 = G_1 G_2 . \mathcal{T}_1'\mathcal{T}_2 \Big\} \subset \{G\}\otimes\{\mathcal{T}\},$
$(G_1\mathcal{T}_1)^{-1} = \mathcal{T}_1^{-1}G_1^{-1} = \mathcal{T}'G' = G'\mathcal{T}'' \Big\}$ $\qquad\qquad (9)$

on bearing in mind the coupling relation (11), Chapter 8. An easy corollary is that this group can be re-demarcated into the form (4), which means that each method of constructing the crystal implies the other. In particular, the translational repetition of the motif pattern produces a crystal structure which shares the pattern's rotational symmetry.

The translation group $\{\mathcal{T}\}$ plays a subsidiary role to that of $\{G\}$ in the space group $\{G\}\otimes\{\mathcal{T}\}$. More precisely, there exists a homomorphic correspondence

$$\{G\}\otimes\{\mathcal{T}\} \to \{G\}, \tag{10}$$

which summarises the correspondence

$$G_1 \mathscr{T}_1 . G_2 \mathscr{T}_2 = G_3 \mathscr{T}_3 \rightarrow G_1 G_2 = G_3$$
$$(G_1 \mathscr{T}_1)^{-1} = G' \mathscr{T}'' \rightarrow G_1{}^{-1} = G'$$

between space-group and point-group equations. According to the terminology of abstract group theory, $\{\mathscr{T}\}$ is an invariant subgroup of $\{G\} \otimes \{\mathscr{T}\}$, so defining a factor or quotient group isomorphic with $\{G\}$, i.e.

$$\frac{\{G\} \otimes \{\mathscr{T}\}}{\{\mathscr{T}\}} \Leftrightarrow \{G\}. \tag{11}$$

Our analysis may be readily extended to the centred translation groups $\{\mathscr{I}\}$, $\{\mathscr{F}\}$, $\{\mathscr{E}\}$, $\{\mathscr{R}\}$ on bearing in mind the supplementary coupling conditions (12)–(15), Chapter 8. By virtue of these conditions, we may simply replace $\{\mathscr{T}\}$ by $\{\mathscr{I}\}$, etc., in (4), (8), (10), (11) above.

9.2 CLASSIFICATION OF SYMMORPHIC SPACE GROUPS

It appears that the point group $\{G\}$ has been used above in two different senses. Thus, $\{G\}$ in (11), Chapter 8, refers to the underlying space lattice, which always has the holohedral symmetry of the crystal system in question. However, $\{G\}$ in (1) *et seq.* above refers to the motif pattern, which could have a lower symmetry than this. For instance, the primitive cubic lattice has a holohedral symmetry $\frac{4}{m}\bar{3}\frac{2}{m}$, which allows the lower cubic symmetries 23, $\frac{2}{m}\bar{3}$, 432, $\bar{4}3m$. Of course no inconsistency arises in these two usages, since any subgroup of $\{G\}$ satisfies (11), Chap. 8, if $\{G\}$ itself does.

Accordingly, there exist five symmorphic space groups associated with the primitive cubic lattice, viz. those symbolised $P23$, $P\frac{2}{m}\bar{3}$, $P432$, $P\bar{4}3m$, $P\frac{4}{m}\bar{3}\frac{2}{m}$. Corresponding to these, there exist the symmorphic space groups $I_{23}, \ldots$ and $F_{23}, \ldots$ associated respectively with the body-centred cubic and face-centred cubic lattices. So proceeding for all the crystal systems, we obtain the sixty-six symmorphic space groups listed in Table 9.1, it being understood that the relevant lattice-type symbol P, I, F, C or R prefixes the point-group symbol.

9.3 AMBIGUITIES OF SETTING

As regards the setting of the motif unit relative to the cell edges, an ambiguity arises in the following seven cases:

$$C2mm; \ P\bar{4}2m, \ I\bar{4}2m; \ P32, \ P3m, \ P\frac{3}{m}m2, \ P\bar{3}\frac{2}{m}. \tag{11}$$

Table 9.1

Enumeration of the sixty-six Bravais space groups. The lattice-type symbols are defined in Fig. 6.13.

Crystal system	Lattice type	Point groups	Number of space groups
cubic	P, I, F	$23, \dfrac{2}{m}\bar{3}$	$3 \times 5 = 15$
		$432, \bar{4}3m$	
		$\dfrac{4}{m}\bar{3}\dfrac{2}{m}$	
tetragonal	P, I	$4, \bar{4}, \dfrac{4}{m}$	$2 \times 7 = 14$
		$422, 4mm, \bar{4}2m$	
		$\dfrac{4}{m}\dfrac{2}{m}\dfrac{2}{m}$	
orthorhombic	P, I, F, C	$222, 2mm$	$4 \times 3 = 12$
		$\dfrac{2}{m}\dfrac{2}{m}\dfrac{2}{m}$	
monoclinic	P, I	$2, m, \dfrac{2}{m}$	$2 \times 3 = 6$
triclinic	P	$1, \bar{1}$	$1 \times 2 = 2$
hexagonal	P, R	$3, 32, 3m$	$2 \times 5 = 10$
		$\bar{3}, \bar{3}\dfrac{2}{m}$	
	P	$6, \bar{6}\left(=\dfrac{3}{m}\right), \dfrac{6}{m}$	$1 \times 7 = 7$
		$6mm, 622, \bar{6}2m\left(=\dfrac{3}{m}2m\right)$	
		$\dfrac{6}{m}\dfrac{2}{m}\dfrac{2}{m}$	
			$\overline{66}$

Thus, C2mm indicates an end-centred orthorhombic cell where [001] functions as a 2-fold symmetry axis, leaving open two distinct possibilities for the centred face, namely

(a) it is (001), i.e. it lies perpendicular to [001], this setting being designated C2mm;
(b) it is either (100) or (010), i.e. it contains [001], this setting being designated A2mm or B2mm (Fig. 9.1).

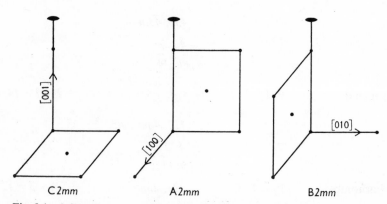

C2mm   A2mm   B2mm

Fig. 9.1—Orientations of a 2-fold axis lying along [001], relative to the centred face of an orthorhombic cell. The orientations A2mm and B2mm are equivalent.

No such ambiguity of course attaches to C222 or C$\dfrac{2}{m}\dfrac{2}{m}\dfrac{2}{m}$, or to any of the other orthorhombic space groups. The next ambiguity arises with P$\bar{4}$2m, for this indicates a primitive tetragonal cell where [001] functions as a $\bar{4}$-fold symmetry axis, leaving open two distinct possibilities for the secondary 2-fold axes, viz.

(a) they are [100] and [010], in which case the mirror planes are (110) and (1$\bar{1}$0), this setting being designated P$\bar{4}$2m;
(b) they are [110] and [1$\bar{1}$0], in which case the mirror planes are (100) and (010), this setting being designated P$\bar{4}$m2 (Fig. 9.2).

Similar considerations hold for I$\bar{4}$2m, I$\bar{4}$m2. Finally, P32 implies a primitive hexagonal cell where [001] functions as a 3-fold symmetry axis, leaving open two distinct possibilities for the secondary 2-fold axes, viz.

(a) they are [100], [010], [1$\bar{1}$0] respectively, this setting being designated P321;
(b) they fall half-way between the preceding directions as indicated in Fig. 9.3, this setting being designated P312.

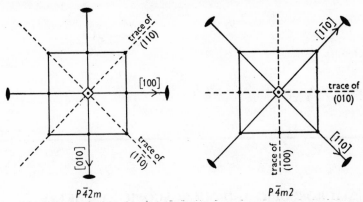

Fig. 9.2—Orientations of secondary 2-fold axes and of accompanying axial mirror planes, relative to a square net. Note that the axial planes passing through [100], [010], [$\bar{1}\bar{1}$0], are (010), (100), (1$\bar{1}$0) respectively.

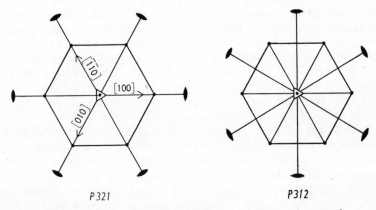

Fig. 9.3—Orientations of secondary 2-fold axes relative to a hexagonal net.

Similar considerations hold for P$3m$, P$\dfrac{3}{m}m2$, P$\bar{3}\dfrac{2}{m}$. Accordingly, we rewrite the space groups (11) as

$$C2mm;\ \text{P}\bar{4}2m,\ \text{I}\bar{4}2m;\ \text{P}321,\ \text{P}3m1,\ \text{P}\frac{3}{m}m2,\ \text{P}\bar{3}\frac{2}{m}1, \tag{12}$$

and supplement them by their partners

$$A2mm;\ \text{P}\bar{4}m2,\ \text{I}\bar{4}m2;\ \text{P}312,\ \text{P}31m,\ \text{P}\frac{3}{m}2m,\ \text{P}\bar{3}1\frac{2}{m}, \tag{13}$$

thereby bringing the total number of Bravais space groups to 73.

Chapter 10

Screw Axes

10.1 MACROSCOPIC AND MICROSCOPIC SYMMETRY

The essential macroscopic symmetry of a crystal may be determined directly from morphological and goniometric evidence. This symmetry always conforms to one of the thirty-two point-groups, a property which would be ensured by the microscopic model of Chapter 9. For instance, the microscopic point-group symmetry *432*, on a primitive cubic lattice, implies a macroscopic symmetry *432*. However, the converse does not necessarily hold. Thus, given a crystal with macroscopic symmetry *432*, we may infer that the underlying space lattice must be primitive, or body-centred, or face-centred cubic. But what can be inferred concerning the microscopic point-group symmetry? An immediately obvious possibility is that this should also conform to *432*. However, in the years following Bravais, microscopic symmetry elements were extended from rotation axes and mirror planes to screw axes and glide planes. A screw operator consists of a rotation coupled with an admissible translation, but the latter gets suppressed at the macroscopic level so resulting in a pure rotation axis at the macroscopic level. Similarly, a macroscopic mirror plane could result from a microscopic glide plane. These extensions imply that a given macroscopic symmetry may be consistent with several distinct microscopic symmetries, each of which is defined mathematically by a space group. The simplest non-Bravais space groups are those involving only screw operators, and we present the theory of these below.

Corresponding to the pure rotation operator C, where $C^n = I$, we construct the screw operator $\mathscr{C}^f C$; $0 < f < 1$. This composite operator signifies a rotation through $2\pi/n$ about [001] followed by a translation through the vector $<0, 0, fc>$. Clearly $\mathscr{C}^f C = C\mathscr{C}^f$ since the translation has the same direction as the axis of rotation (Fig. 10.1). If $\mathscr{C}^f C$ belongs to a space group, then so must its powers

$$\mathscr{C}^f C, \quad \mathscr{C}^{2f} C^2, \ldots, \quad \mathscr{C}^{nf},$$

$$\mathscr{C}^{nf+f} C, \quad \mathscr{C}^{nf+2f} C^2, \ldots, \quad \mathscr{C}^{2nf}, \tag{1}$$

$$\mathscr{C}^{2nf+f} C, \ldots$$

which are seen to include the pure translation operators $\mathscr{C}^{nf}$, $\mathscr{C}^{2nf}$, ...

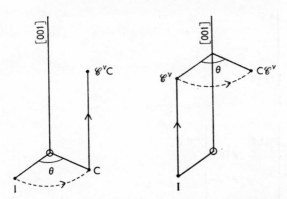

Fig. 10.1—Each point is symbolized by the net operations which produces it from I. Evidently $\mathscr{C}^\nu C$, $C\mathscr{C}^\nu$ define the same point.

However, the pure translation operators for the [001] direction can only be $\mathscr{C}$, $\mathscr{C}^2$, ... as is required by the crystal's translational symmetry. Accordingly $\mathscr{C}^{nf} = \mathscr{C}$, showing that an admissible screw operator is $\mathscr{C}^{1/n}C$. The other allowed screw operators associated with C are $\mathscr{C}^{2/n}C$, $\mathscr{C}^{3/n}C$, ... $\mathscr{C}^{(n-1)/n}C$, since each of these generates powers which eventually coincide with pure translational operators. Consequently, associated with the rotation axes $2, 3, 4, 6,$ we have the screw axes

$$2_1; \qquad 3_1, 3_2; \qquad 4_1, 4_2, 4_3; \qquad 6_1, 6_2, 6_3, 6_4, 6_5 \qquad (2)$$

utilising an obvious notation. These repeat a point in space as indicated in Fig. 10.2.

If $f = 0$ above, we obtain the pure rotation operator C. If $f = 1$, we obtain the operator $\mathscr{C}C$, which may be formally regarded as a screw operator with powers

$$\left.\begin{array}{l} \mathscr{C}C, \qquad \mathscr{C}^2C^2, \ldots \ldots, \mathscr{C}^n \\ \mathscr{C}^{n+1}C, \quad \mathscr{C}^{n+2}C, \ldots \ldots, \mathscr{C}^{2n} \\ \mathscr{C}^{2n+1}C, \ldots \end{array}\right\}. \qquad (3)$$

However nothing is gained from this point of view, since all these operators are included in the Bravais space group

$$\{I, C, C^2, \ldots C^{n-1}\} \otimes \{\mathscr{T}\}. \qquad (4)$$

10.2 CYCLIC GROUPOIDS

By comparison with the cyclic group

$$\{n\} = \{I, C, C^2, \ldots, C^{n-1}\}; \quad C^n = I, \qquad (5)$$

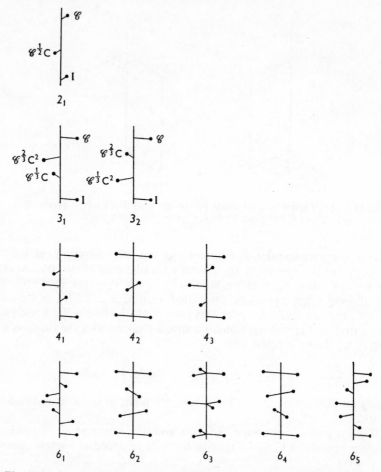

Fig. 10.2—The repetition of a point by screw axes. For 3_2 each point is marked by its equivalent reduced operator. Similar considerations hold for 4_2, 4_3, 6_2, 6_3, ... as can be seen by reference to Table 10.1.

which describes a n-fold rotation, we construct the set

$$\{n_1\} = \{I, \mathscr{C}^{1/n}C, \mathscr{C}^{2/n}C^2, \ldots, \mathscr{C}^{(n-1)/n}C^{n-1}\}; \quad C^n = I, \tag{6}$$

which describes a n-fold screw. This set does not constitute a group, since

$$(\mathscr{C}^{1/n}C)^n = \mathscr{C}^{n/n}C^n = \mathscr{C} \notin \{n_1\},$$

but a close correspondence evidently exists between $\{n_1\}$ and $\{n\}$. Thus, corresponding to the equations

$$C^p . C^q = C^{p+q} = C^{vn+r} = C^r; \quad 0 < p, \, q, \, r < n \atop v = 0, \, 1 \Bigg\} \qquad (7)$$

$$(C^p)^{-1} = C^{-p} = C^{n-p}$$

appertaining to $\{n\}$, we have the equations

$$\mathscr{C}^{p/n}C^p . \mathscr{C}^{q/n}C^q = \mathscr{C}^{(p+q)/n} C^{p+q} = \mathscr{C}^v . \mathscr{C}^{r/n} C^r \atop (\mathscr{C}^{p/n} C^p)^{-1} = C^{-p} \mathscr{C}^{-p/n} = \mathscr{C}^{-1} . \mathscr{C}^{(n-p)/n}C^{n-p} \Bigg\} \qquad (8)$$

appertaining to $\{n_1\}$. Apart from the presence of $\mathscr{C}^{-1}$, $\mathscr{C}^v$ in (8), both of which are lattice translation operators, these equations are isomorphic with the equations (7). It is convenient to re-write (8) as

$$\mathscr{C}^{p/n}C^p . \mathscr{C}^{q/n}C^q = \mathscr{C}^{(p+q)/n} C^{p+q} \Rightarrow \mathscr{C}^{r/n}C^r; \quad p+q = vn+r \atop (\mathscr{C}^{p/n}C^p)^{-1} = C^{-p} \mathscr{C}^{-p/n} \Rightarrow \mathscr{C}^{(n-p)/n} C^{n-p} \Bigg\}, \qquad (8a)$$

where the symbol $\Rightarrow$ indicates equality up to an appropriate member of $\{\mathscr{T}\}$. All these particular equivalences may be summarised by writing

$$\{n_1\} \Leftrightarrow \{n\} \text{ modulo } \{\mathscr{T}\}, \qquad (9)$$

i.e. $\{n_1\}$ is isomorphic with the cyclic group of order n apart from factors which belong to $\{\mathscr{T}\}$. Slightly adapting the terminology of Buerger [6], we say that $\{n_1\}$ is isogonal with $\{n\}$, though this also expresses the related equivalence (12) below.

It follows from (8), and from the coupling relation

$$\mathscr{C}^{p/n}C^p\mathscr{T}_1 . \mathscr{T}_2 . (\mathscr{C}^{p/n} C^p\mathscr{T}_1)^{-1} = C^p\mathscr{T}_2C^{-p} = \mathscr{T}_2', \qquad (10)$$

that the infinite set of operators

$$\{n_1\} \otimes \{\mathscr{T}\} = \{\mathscr{T}\}, \quad \mathscr{C}^{1/n}C\{\mathscr{T}\}, \dots, \mathscr{C}^{(n-1)/n}C^{n-1}\{\mathscr{T}\} \qquad (11)$$

forms a group containing $\{\mathscr{T}\}$ as an invariant subgroup. This is the simplest example of a non-Bravais space group. By comparison with (10) *et al*, Chap. 9, there exists a homomorphic correspondence

$$\{n_1\} \otimes \{\mathscr{T}\} \rightarrow \{n\} \qquad (12)$$

which in effect provides an equivalent statement to (9) above. Also, following (11), Chap. 9, we may write

$$\frac{\{n_1\} \otimes \{\mathscr{T}\}}{\{\mathscr{T}\}} \Leftrightarrow \{n\}, \qquad (13)$$

signifying the existence of a factor or quotient group isomorphic with $\{n\}$, but the statement (13) adds nothing to (12).

The space group $\{n_1\} \otimes \{\mathcal{T}\}$ produces the same macroscopic symmetry as does $\{n\} \otimes \{\mathcal{T}\}$, i.e. one characterised by the point-group $\{n\}$. At the microscopic level, however, $\{n_1\} \otimes \{\mathcal{T}\}$ produces a screw pattern whereas $\{n\} \otimes \{\mathcal{T}\}$ produces a pure rotation pattern. This aspect will now be pursued.

10.3 SIGNIFICANCE OF $\{n_1\} \otimes \{\mathcal{T}\} = \{\mathcal{T}\} \otimes \{n_1\}$

All the equivalent atoms of a crystal may be built up from any one of them by the operators $\{\mathcal{T}\} \otimes \{n_1\}$. Thus, following the procedure of section 9.1, we first place an atom at $\mathbf{R}$ relative to 0, where $\mathbf{R}$ lies within any of the eight unit cells surrounding 0. Equivalent atoms then appear at the n distinct positions $\{n_1\}\mathbf{R}$, so forming an elementary screw pattern about an axis through 0. This pattern lies entirely within the eight unit cells surrounding 0. Operating upon these positions by a translation operator $\mathcal{T}$, we generate further equivalent atoms at $\mathcal{T}\{n_1\}\mathbf{R}$, so producing a replica of the original screw pattern translated through a lattice vector $\mathbf{t}$. So proceeding for every choice of $\mathcal{T}$ as given by (4), Chap. 8, we generate all the equivalent screw patterns—and hence all the equivalent atoms of the crystal—by the infinite set of operators $\{\mathcal{T}\} \otimes \{n_1\}$. This is the space group of the crystal constructed so as to emphasise its translational symmetry. It fails, however, to expose the crystal's screw symmetry as a whole with respect to the axis through 0.

An alternative procedure is to place an atom at $\mathbf{R}$ as before, and then generate equivalent atoms at $\{\mathcal{T}\}\mathbf{R}$, i.e. at positions $\mathbf{R}$ relative to each lattice point. Operating upon these positions by $\mathscr{C}^{p/n}C^p$, with reference to a screw axis passing through 0, we generate further equivalent atoms at $\mathscr{C}^{p/n}C^p\{\mathcal{T}\}\mathbf{R}$. So proceeding for all the operators of $\{n_1\}$ as appear in (6) above, we generate all the equivalent atoms of the crystal by the infinite set of operators $\{n_1\} \otimes \{\mathcal{T}\}$. This is the space group of the crystal, constructed so as to emphasise its screw symmetry as a whole about the axis through 0. It fails, however, to expose the elementary screw pattern associated with each lattice point. The two methods of construction are of course equivalent, since $\{n_1\} \otimes \{\mathcal{T}\} = \{\mathcal{T}\} \otimes \{n_1\}$ by virtue of (10) above.

The other cyclic groupoids isogonal with $\{n\}$ are

$$\{n_2\}, \ldots, \{n_{n-1}\}, \text{ where}$$
$$\{n_2\} = \{I, \mathscr{C}^{2/n}C, \mathscr{C}^{4/n}C^2, \ldots, \mathscr{C}^{2(n-1)/n}C^{n-1}\}, \text{ etc.} \tag{14}$$

It will be noted that $\{n_2\}$ contains operators of the form

$$\mathscr{C}^{2p/n}C^p; \quad 2 > \frac{2p}{n} \geqslant 1, \tag{15}$$

which involve a translational component reducible to

$$\mathscr{C}^{-1}\,\mathscr{C}^{2p/n}=\mathscr{C}^{(2p-n)/n};\quad 1>\frac{2p-n}{n}\geqslant 0 \qquad (16)$$

This reduction does not affect the space group $\{n_2\}\otimes\{\mathscr{T}\}$, and it allows a more convenient geometrical picture of the screw operation. Accordingly, we henceforth regard $\{n_2\}$ as the reduced cyclic groupoid associated with (14) above, and similarly for $\{n_3\}$, $\{n_4\}$, $\{n_6\}$ as given in Table 10.1 and displayed geometrically in Fig. 10.2.

Table 10.1

$\{3_1\}=I \quad \mathscr{C}^{\frac{1}{3}}C \quad \mathscr{C}^{\frac{2}{3}}C^2$

$\{3_2\}=I \quad \mathscr{C}^{\frac{2}{3}}C \quad \mathscr{C}^{\frac{4}{3}}C^2$

$\{3_2\}=I \quad \mathscr{C}^{\frac{2}{3}}C \quad \mathscr{C}^{\frac{1}{3}}C^2$

$\{4_1\}=I \quad \mathscr{C}^{\frac{1}{4}}C \quad \mathscr{C}^{\frac{2}{4}}C^2 \quad \mathscr{C}^{\frac{3}{4}}C^3$

$\{4_2\}=I \quad \mathscr{C}^{\frac{2}{4}}C \quad \mathscr{C}^{\frac{4}{4}}C^2 \quad \mathscr{C}^{\frac{6}{4}}C^3$

$\{4_2\}=I \quad \mathscr{C}^{\frac{1}{2}}C \quad C^2 \quad \mathscr{C}^{\frac{1}{2}}C^2$

$\{4_3\}=I \quad \mathscr{C}^{\frac{3}{4}}C \quad \mathscr{C}^{\frac{6}{4}}C^2 \quad \mathscr{C}^{\frac{9}{4}}C^3$

$\{4_3\}=I \quad \mathscr{C}^{\frac{3}{4}}C \quad \mathscr{C}^{\frac{1}{2}}C^2 \quad \mathscr{C}^{\frac{1}{4}}C^3$

$\{6_1\}=I \quad \mathscr{C}^{\frac{1}{6}}C \quad \mathscr{C}^{\frac{2}{6}}C^2 \quad \mathscr{C}^{\frac{3}{6}}C^3 \quad \mathscr{C}^{\frac{4}{6}}C^4 \quad \mathscr{C}^{\frac{5}{6}}C^5$

$\{6_2\}=I \quad \mathscr{C}^{\frac{2}{6}}C \quad \mathscr{C}^{\frac{4}{6}}C^2 \quad \mathscr{C}^{\frac{6}{6}}C^3 \quad \mathscr{C}^{\frac{8}{6}}C^4 \quad \mathscr{C}^{\frac{10}{6}}C^5$

$\{6_2\}=I \quad \mathscr{C}^{\frac{1}{3}}C \quad \mathscr{C}^{\frac{2}{3}}C^2 \quad C^3 \quad \mathscr{C}^{\frac{1}{3}}C^4 \quad \mathscr{C}^{\frac{2}{3}}C^5$

$\{6_3\}=I \quad \mathscr{C}^{\frac{3}{6}}C \quad \mathscr{C}^{\frac{6}{6}}C^2 \quad \mathscr{C}^{\frac{9}{6}}C^3 \quad \mathscr{C}^{\frac{12}{6}}C^4 \quad \mathscr{C}^{\frac{15}{6}}C^5$

$\{6_3\}=I \quad \mathscr{C}^{\frac{1}{2}}C \quad C^2 \quad \mathscr{C}^{\frac{1}{2}}C^3 \quad C^4 \quad \mathscr{C}^{\frac{1}{2}}C^5$

$\{6_4\}=I \quad \mathscr{C}^{\frac{4}{6}}C \quad \mathscr{C}^{\frac{8}{6}}C^2 \quad \mathscr{C}^{\frac{12}{6}}C^3 \quad \mathscr{C}^{\frac{16}{6}}C^4 \quad \mathscr{C}^{\frac{20}{6}}C^5$

$\{6_4\}=I \quad \mathscr{C}^{\frac{2}{3}}C \quad \mathscr{C}^{\frac{1}{3}}C^2 \quad C^3 \quad \mathscr{C}^{\frac{2}{3}}C^4 \quad \mathscr{C}^{\frac{1}{3}}C^5$

$\{6_5\}=I \quad \mathscr{C}^{\frac{5}{6}}C \quad \mathscr{C}^{\frac{10}{6}}C^2 \quad \mathscr{C}^{\frac{15}{6}}C^3 \quad \mathscr{C}^{\frac{20}{6}}C^4 \quad \mathscr{C}^{\frac{25}{6}}C^5$

$\{6_5\}=I \quad \mathscr{C}^{\frac{5}{6}}C \quad \mathscr{C}^{\frac{4}{6}}C^2 \quad \mathscr{C}^{\frac{3}{6}}C^3 \quad \mathscr{C}^{\frac{2}{6}}C^4 \quad \mathscr{C}^{\frac{1}{6}}C^5$

Arrows ⌐ connect screws related as left to right.

Arrows ⊂ indicate reduced versions of appropriate groupoids.

Chapter 11

Principal and Secondary Screw Axes

11.1 THEORY OF COMPOSITE OPERATORS

The composite operator $\mathscr{A}^\lambda\mathscr{B}^\mu\mathscr{C}^\nu\,(hkl)^{1/2}$ signifies a rotation through π about $[hkl]$ followed by a translation $<\lambda,\mu,\nu>$. If this lies transverse to the axis, i.e. if $[\lambda\mu\nu].[hkl] = 0$, there results (Appendix 6) a pure rotation through π about an axis parallel to the original and separated from it by the translation $\frac{1}{2}<\lambda,\mu,\nu>$. Since two successive rotations through π yield the identity, we obtain

$$\mathscr{A}^\lambda\mathscr{B}^\mu\mathscr{C}^\nu(hkl)^{1/2}.\mathscr{A}^\lambda\mathscr{B}^\mu\mathscr{C}^\nu(hkl)^{1/2} = I,$$

i.e.

$$\mathscr{A}^\lambda\mathscr{B}^\mu\mathscr{C}^\nu(hkl)^{1/2} = (hkl)^{1/2}\mathscr{A}^{-\lambda}\mathscr{B}^{-\mu}\mathscr{C}^{-\nu}, \tag{1}$$

on bearing in mind $(hkl)^{1/2} = (hkl)^{-1/2}$, which shows that a rotation does not commute with a transverse translation. An equivalent statement to (1) is

$$(hkl)^{1/2}\mathscr{A}^\lambda\mathscr{B}^\mu\mathscr{C}^\nu(hkl)^{-1/2} = \mathscr{A}^{-\lambda}\mathscr{B}^{-\mu}\mathscr{C}^{-\nu}, \tag{2}$$

which is seen to be a particular case of the coupling relation (11), Chapter 8. So far λ,μ,ν may be any rational numbers. If they are integers, our analysis formally still holds but then of course (adopting the symbolism of (8a), Chapter 10)

$$\mathscr{A}^\lambda\mathscr{B}^\mu\mathscr{C}^\nu(hkl)^{1/n} \Rightarrow (hkl)^{1/n}. \tag{3}$$

For a rotation through $2\pi/n$ about $[hkl]$, the separation vector $\frac{1}{2}<\lambda,\mu,\nu>$ must be supplemented by an additional separation vector, at right angles to both $[hkl]$ and $[\lambda\mu\nu]$ as specified in Appendix 6. It is sometimes convenient to signify rotations by writing

$$\mathscr{A}^\lambda\mathscr{B}^\mu\mathscr{C}^\nu(hkl)^{1/2} = \left(\frac{\lambda}{2}\frac{\mu}{2}\frac{\nu}{2}/hkl\right)^{1/2}, \tag{4}$$

$$\mathscr{A}^\lambda \mathscr{B}^\mu \mathscr{C}^\nu (hkl)^{1/n} = \left(\frac{\lambda}{2}\frac{\mu}{2}\frac{\nu}{2}/hkl\right)_+^{1/n}, \tag{5}$$

where the subscript $+$ in (5) indicates a supplementary translation as compared with (4).

If $<\lambda, \mu, \nu>$ lies parallel to the axis, i.e. $<\lambda, \mu, \nu> = \frac{1}{n}<h, k, l>$, then we obtain a screw operator conveniently written

$$[hkl]^{1/n}(hkl)^{1/n}. \tag{6}$$

Here the translation and rotation commute, yielding the powers

$$[hkl]^{r/n}(hkl)^{r/n}; \quad r = 1, 2, \ldots n-1. \tag{7}$$

More generally, if

$$<\lambda, \mu, \nu> = <f, g, h> + \frac{1}{n}<h, k, l>; \quad [fgh].[hkl] = 0, \tag{8}$$

where f, g, h are rational, then the screw operator appears as

$$[hkl]^{1/n}\left(\frac{f}{2}\frac{g}{2}\frac{h}{2}/hkl\right)_+^{1/n}. \tag{9}$$

If we are unable to provide a rational breakdown (8), then the composite operator $\mathscr{A}^\lambda \mathscr{B}^\mu \mathscr{C}^\nu (hkl)^{1/n}$ does not qualify as a space-group operator. Consequently, any set which includes it would not be admissible as a crystallographic groupoid.

11.2 DIHEDRAL GROUPOIDS: PRINCIPAL SCREW AXES
Slightly adapting the symbolism of Chapter 2, we now write

$$\{n2\} = \begin{Bmatrix} I, C, & \ldots, & C^{n-1} \\ A, AC, & \ldots, & AC^{n-1} \end{Bmatrix}; \quad A^2 = C^n = (AC)^2 = I \tag{10}$$
$$AC^r = C^{n-r}A$$

where

$$C = (001)^{1/n}, \quad A = (100)^{1/2}, \quad AC^r = (h,k,0)^{1/2}.$$

By analogy with (6), Chapter 10, this suggests the construction of a groupoid

$$\{n_12\} = \begin{cases} I, \mathscr{C}^{1/n}C, & \ldots\ldots, \mathscr{C}^{(n-1)/n}C \\ A, A\mathscr{C}^{1/n}C, & \ldots\ldots, A\mathscr{C}^{(n-1)/n}C \end{cases},$$

$$= \begin{cases} I, \mathscr{C}^{1/n}C, & \ldots\ldots, \mathscr{C}^{(n-1)/n}C \\ A, \mathscr{C}^{-1/n}AC, & \ldots\ldots, \mathscr{C}^{-(n-1)/n}AC^{n-1} \end{cases}, \tag{11}$$

since $A\mathscr{C}^{r/n}A^{-1} = \mathscr{C}^{-r/n}$. It may be readily verified that

$$\{n_12\} \Leftrightarrow \{n2\} \bmod \{\mathscr{T}\}, \tag{12}$$

e.g. the product in (10)

$$AC^{n-1} \cdot C^{n-2} = AC^{2n-3} = AC^{n-3}$$

corresponds with the product in (11)

$$\mathscr{C}^{-(n-1)/n}AC^{n-1} \cdot \mathscr{C}^{(n-2)/n}C^{n-2} = \mathscr{C}^{-(2n-3)/n}AC^{2n-3} \Rightarrow$$
$$\mathscr{C}^{-(n-3)/n}AC^{n-3}.$$

Also

$$(\mathscr{C}^{-r/n}AC^r)^2 = I; \quad r = 0, 1, 2, \ldots, n-1$$

as may be proved by the methods of Appendix 1 or more directly by noting that

$$\mathscr{C}^{-r/n}AC^r \cdot \mathscr{C}^{-r/n}AC^r = \mathscr{C}^{-r/n} \cdot AC^r\mathscr{C}^{-r/n} \cdot AC^r = \mathscr{C}^{-r/n} \cdot \mathscr{C}^{r/n}AC^r \cdot AC^r$$
$$= \mathscr{C}^{-r/n}\mathscr{C}^{+r/n} \cdot (AC^r)^2 = I,$$

which shows that the composite operator $\mathscr{C}^{-r/n}AC^r$ signifies a rotation through π. This conclusion is, of course, consistent with the identification

$$\mathscr{C}^{-r/n}AC^r = \mathscr{C}^{-r/n}(h_r k_r 0)^{1/2} = (0\ 0\ \frac{\bar{r}}{2n}/h_r k_r\ 0)^{1/2}$$

which follows on noting that [001].$[h_r k_r 0] = 0$ for all cell axes other than triclinic or rhombohedral. Accordingly, $\{n_12\}$ defines a n-fold principal screw axis of symmetry, combined with n secondary rotation axes spaced by successive screw operations. Similar considerations hold for $\{n_22\}$, $\{n_32\}$, ... Remembering that $\{n2\}$ expands into $\{n22\}$ for $n = 2, 4, 6$ and that $\{32\}$ appears as both $\{321\}$ and $\{312\}$, we arrive at the relevant space groups listed in Table 11.1.

Table 11·1

Crystal system	Lattice type	Groupoid				
monoclinic	P	2_1				
orthorhombic	P	2_122	2_12_12	$2_12_12_1$		
tetragonal	P	4_1	4_2	4_3		
		4_122	4_222	4_322		
		4_12_12	4_22_12	4_32_12	42_12	
hexagonal	P	6_1	6_2	6_3	6_4	6_5
		6_122	6_222	6_322	6_422	6_522
	P, R	3_1	3_2			
		3_121	3_221			
		3_112	3_212			
cubic	P	2_13				
		4_132	4_232	4_332		

11.3 DIHEDRAL GROUPOIDS: SECONDARY SCREW AXES

We now enquire whether some or all of the secondary rotation axes can be replaced by 2-fold screw axes. An obvious mathematical step would be to replace A by $\mathscr{A}^{1/2}A$ in (10), so yielding the set

$$\left\{ \begin{array}{cccc} I, & C, \ldots\ldots\ldots, & C^{n-1} \\ \mathscr{A}^{1/2}A, & \mathscr{A}^{1/2}AC, \ldots\ldots, & \mathscr{A}^{1/2}AC^{n-1} \end{array} \right\}. \tag{13}$$

However, apart from $\mathscr{A}^{1/2}A$, no operator of the form $\mathscr{A}^{1/2}AC^r$ carries geometrical significance either as a pure rotation or screw:

$$\mathscr{A}^{1/2}AC^r = \mathscr{A}^{1/2}(h_r k_r 0)^{1/2}; \quad [100].[h_r k_r 0] \neq 0$$
$$[100] \wedge [h_r k_r 0] \neq 0.$$

An exception occurs when $n = 2$, since then:

$$\mathscr{A}^{1/2}A = [100]^{1/2}(100)^{1/2} \quad \text{i.e. a screw operator,}$$

$$\mathscr{A}^{1/2}AC = \mathscr{A}^{1/2}(010)^{1/2} = (\tfrac{1}{4}00/010)^{1/2} \quad \text{i.e. a rotation.}$$

This allows the construction of a dihedral groupoid

$$\{22_12\} = \{I, C, \mathscr{A}^{1/2}A, \mathscr{A}^{1/2}AC\}; \quad A^2 = C^2 = (AC)^2 = I; \quad AC = CA \tag{14}$$

with the property $\{22_12\} \Leftrightarrow \{222\} \bmod \{\mathscr{T}\}$.

But clearly $\{22_12\}$ must be equivalent to $\{2_122\}$, i.e. putting $n = 2$ in (10), which shows that nothing fruitful emerges from the set (13).

A preferable approach is to replace A by $\mathscr{A}^{1/2}\mathscr{B}^{1/2}A$ in (10), which yields admissible new groupoids for the cases $n = 2, 4$. Thus, when $n = 2$, we obtain

$$\{22_12_1\} = \{I, C, \mathscr{A}^{1/2}\mathscr{B}^{1/2}A, \ \mathscr{A}^{1/2}\mathscr{B}^{1/2}AC\}; \quad A^2 = C^2 = (AC)^2 = I$$
$$AC = CA \tag{15}$$

where

$$\left.\begin{array}{l} \mathscr{A}^{1/2}\mathscr{B}^{1/2}A = \mathscr{A}^{1/2}\mathscr{B}^{1/2}(100)^{1/2} = [100]^{1/2}(0\tfrac{1}{4}0/100)^{1/2} \\ \mathscr{A}^{1/2}\mathscr{B}^{1/2}AC = \mathscr{A}^{1/2}\mathscr{B}^{1/2}(010)^{1/2} = [010]^{1/2}(\tfrac{1}{4}00/010)^{1/2} \end{array}\right\} . \tag{16}$$

Each of these is seen to be a 2-fold screw operator, so accounting for the symbolism $\{22_12_1\}$, and it may be readily proved that $\{22_12_1\} \Leftrightarrow \{222\}$ mod $\{\mathscr{T}\}$ on bearing in mind (Appendix 1)

$$C\mathscr{A}^\lambda\mathscr{B}^\mu C^{-1} = \mathscr{A}^{-\lambda}\mathscr{B}^{-\mu}, \quad A\mathscr{A}^\lambda\mathscr{B}^\mu A^{-1} = \mathscr{A}^\lambda\mathscr{B}^{-\mu}; \quad C^2 = I. \tag{17}$$

When $n = 4$ we obtain

$$\{42_12\} = \left\{\begin{array}{cccc} I, & C, & C^2, & C^3 \\ \mathscr{A}^{1/2}\mathscr{B}^{1/2}A, & \mathscr{A}^{1/2}\mathscr{B}^{1/2}AC, & \mathscr{A}^{1/2}\mathscr{B}^{1/2}AC^2, & \mathscr{A}^{1/2}\mathscr{B}^{1/2}AC^3 \end{array}\right\}$$
$$A^2 = C^4 = (AC)^2 = I, \ AC^r = C^{4-r}A, \tag{18}$$

where:

$$\left.\begin{array}{l} \mathscr{A}^{1/2}\mathscr{B}^{1/2}A \ \ = \mathscr{A}^{1/2}\mathscr{B}^{1/2}(100)^{1/2} = [100]^{1/2}(0\tfrac{1}{4}0/100)^{1/2} \\ \mathscr{A}^{1/2}\mathscr{B}^{1/2}AC^2 = \mathscr{A}^{1/2}\mathscr{B}^{1/2}(010)^{1/2} = [010]^{1/2}(\tfrac{1}{4}00/010)^{1/2} \end{array}\right\} \text{as in (16)}$$

$$\mathscr{A}^{1/2}\mathscr{B}^{1/2}AC \ = \mathscr{A}^{1/2}\mathscr{B}^{1/2}(1\bar{1}0)^{1/2} = (\tfrac{1}{4}\tfrac{1}{4}0/1\bar{1}0)^{1/2}$$
$$\mathscr{A}^{1/2}\mathscr{B}^{1/2}AC^3 = \mathscr{A}^{1/2}\mathscr{B}^{1/2}(110)^{1/2} = \mathscr{B}.\mathscr{A}^{1/2}\mathscr{B}^{-1/2}(110)^{1/2} \Rightarrow (\tfrac{1}{4}\tfrac{\bar{1}}{4}0/$$
$$110)^{1/2} \text{ mod } \{\mathscr{T}\}.$$

Here the first pair of operators are 2-fold screws and the second pair are interleaving 2-fold rotations, all secondary to the 4-fold principal rotation axis of symmetry, so accounting for the symbolism $\{42_12\}$. It may be proved that $\{42_12\} \Leftrightarrow \{422\}$ mod $\{\mathscr{T}\}$ on bearing in mind (Appendix 1)

$$C\mathscr{A}^\lambda\mathscr{B}^\mu C^{-1} = \mathscr{A}^{-\mu}\mathscr{B}^\lambda, \quad A\mathscr{A}^\lambda\mathscr{B}^\mu A^{-1} = \mathscr{A}^\lambda B^{-\mu}; \quad C^4 = I \tag{19}$$

For $n = 3, 6$ the operator $\mathscr{A}^{1/2}\mathscr{B}^{1/2}A$ carries no geometrical significance, either as a rotation or as a screw, so ruling out the possibility of dihedral groups analogous to (18).

We now enquire whether the principal axis of rotation and the secondary axes may be simultaneously replaced by screws. Only the cases $n=2, 4$ enter into consideration. Replacing C by $\mathscr{C}^{1/2}C$ and A by $\mathscr{A}^{1/2}\mathscr{B}^{1/2}A$ in (10), for $n=2$, yields the set

$$\{I, \mathscr{C}^{1/2}C, \mathscr{A}^{1/2}\mathscr{B}^{1/2}A, \quad \mathscr{A}^{1/2}\mathscr{B}^{1/2}\mathscr{C}^{-1/2}AC\}, \tag{20}$$

where all the operators—apart from I—are 2-fold screws:

$\mathscr{C}^{1/2}C, \quad \mathscr{A}^{1/2}\mathscr{B}^{1/2}A$ have already been covered and

$$\mathscr{A}^{1/2}\mathscr{B}^{1/2}\mathscr{C}^{-1/2}AC = \mathscr{B}^{1/2}.\mathscr{A}^{1/2}\mathscr{C}^{-1/2}(010)^{1/2}$$
$$= \mathscr{B}^{1/2}(\tfrac{1}{4}0\tfrac{\bar 1}{4}/010)^{1/2} = [010]^{1/2}(\tfrac{1}{4}0\tfrac{\bar 1}{4}/010)^{1/2}.$$

However, an equivalent, but more symmetrical, groupoid is

$$\{2_12_12_1\} = \{I, \mathscr{A}^{1/2}\mathscr{B}^{1/2}A, \quad \mathscr{B}^{1/2}\mathscr{C}^{1/2}B, \quad \mathscr{C}^{1/2}\mathscr{A}^{1/2}C\}; \quad B = AC = CA \tag{21}$$
$$A^2 = B^2 = C^2 = I$$

which readily transforms into (20) on writing

$$A = A^*, \quad C = \mathscr{A}^{-1/2}C^*, \quad B = AC = \mathscr{A}^{-1/2}A^*C^*.$$

Of course $\{2_12_12_1\} \Leftrightarrow \{222\} \bmod \{\mathscr{T}\}$.

Similarly, replacing C by $\mathscr{C}^{1/4}C$ and A by $\mathscr{A}^{1/2}\mathscr{B}^{1/2}A$ in (9), for $n=4$, yields

$$\{4_12_12\} = \left\{ \begin{array}{cccc} I , & \mathscr{C}^{1/4}C , & \mathscr{C}^{1/2}C^2 , & \mathscr{C}^{3/4}C^3 \\ \mathscr{A}^{1/2}\mathscr{B}^{1/2}A, & \mathscr{A}^{1/2}\mathscr{B}^{1/2}\mathscr{C}^{-1/4}AC, & \mathscr{A}^{1/2}\mathscr{B}^{1/2}\mathscr{C}^{-1/2}AC^2, \\ & \mathscr{A}^{1/2}\mathscr{B}^{1/2}\mathscr{C}^{-3/4}AC^3 & & \end{array} \right\} \tag{22}$$

where all the secondary operators retain the same significance as in (18):

$$\left. \begin{array}{l} \mathscr{A}^{1/2}\mathscr{B}^{1/2}A = [100]^{1/2}(0\tfrac{1}{4}0/100)^{1/2} \\ \mathscr{A}^{1/2}\mathscr{B}^{1/2}\mathscr{C}^{-1/2}AC^2 = \mathscr{B}^{1/2}\mathscr{A}^{1/2}\mathscr{C}^{-1/2}AC^2 = [010]^{1/2}(\tfrac{1}{4}0\tfrac{\bar 1}{4}/010)^{1/2} \end{array} \right\}$$

$$\left. \begin{array}{l} \mathscr{A}^{1/2}\mathscr{B}^{1/2}\mathscr{C}^{-1/4}AC = (\tfrac{1}{4}\tfrac{1}{4}\tfrac{\bar 1}{8}/1\bar 10)^{1/2} \\ \mathscr{A}^{1/2}\mathscr{B}^{1/2}\mathscr{C}^{-3/4}AC^3 \Rightarrow (\tfrac{1}{4}\tfrac{1}{4}\tfrac{3}{8}/110)^{1/2} \end{array} \right\}.$$

Accordingly, we have a pair of 2-fold screws interleaving a pair of 2-fold rotations, all secondary to the 4-fold principal screw axes of symmetry, so accounting for the symbolism $\{4_12_12\}$.

A similar analysis holds for the groupoids $\{4_22_12\}$, $\{4_32_12\}$ and we note that

$$\{4_12_12\}, \quad \{4_22_12\}, \quad \{4_32_12\} \Leftrightarrow \{422\} \bmod \{\mathscr{T}\}.$$

11.4 CUBIC GROUPOIDS

Starting with the tetrahedral group

$$\{23\} = \{222\} + (111)^{1/3}\{222\} + (111)^{2/3}\{222\} \tag{23}$$

introduced in (3), Chapter 3, we construct the tetrahedral groupoid

$$\{2_13\} = \{2_12_12_1\} + (111)^{1/3}\{2_12_12_1\} + (111)^{2/3}\{2_12_12_1\}, \tag{24}$$

where all the operators in $(111)^{1/3}\{2_12_12_1\}$, $(111)^{2/3}\{2_12_12_1\}$ prove to be 3-fold rotations. Thus, for instance,

$$(111)^{1/3} \cdot \mathscr{A}^{1/2}\mathscr{B}^{1/2}A = (111)^{1/3}\mathscr{A}^{1/2}\mathscr{B}^{1/2}A = \mathscr{B}^{1/2}\mathscr{C}^{1/2}(111)^{1/3}(100)^{1/2}$$
$$= \mathscr{B}^{1/2}\mathscr{C}^{1/2}(\bar{1}\bar{1}1)^{1/3} = (0\tfrac{1}{4}\tfrac{1}{4}/\bar{1}\bar{1}1)^{1/3}_{+}, \tag{25}$$

on bearing in mind

$$\left.\begin{array}{l}(111)^{1/3}\mathscr{A}^{\lambda}\mathscr{B}^{\mu}\mathscr{C}^{\nu}(111)^{-1/3} = \mathscr{A}^{\nu}\mathscr{B}^{\lambda}\mathscr{C}^{\mu}\\(111)^{2/3}\mathscr{A}^{\lambda}\mathscr{B}^{\mu}\mathscr{C}^{\nu}(111)^{-2/3} = \mathscr{A}^{\mu}\mathscr{B}^{\nu}\mathscr{C}^{\lambda}\end{array}\right\} \tag{26}$$

and the products $(111)^{1/3}(100)^{1/2} = (\bar{1}\bar{1}1)^{1/3}$, etc., which enter into $\{23\}$. Alternatively, we may directly compute the product $(111)^{1/3}[100]^{1/2}[010]^{1/2}$ $(100)^{1/2}$, and identify it with $[010]^{1/2}[001]^{1/2}(\bar{1}\bar{1}1)^{1/3}$ following the method of Appendix 1. We find:

$$\left.\begin{array}{ll}(111)^{1/3}\mathscr{A}^{1/2}\mathscr{B}^{1/2}A = (0\tfrac{1}{4}\tfrac{1}{4}/\bar{1}\bar{1}1)^{1/3}_{+}, & (111)^{2/3}\mathscr{A}^{1/2}\mathscr{B}^{1/2}A \Rightarrow \tfrac{1}{4}(\tfrac{1}{4}0\bar{1}/\bar{1}1\bar{1})^{2/3}_{+}\\(111)^{1/3}\mathscr{B}^{1/2}\mathscr{C}^{1/2}B = (\tfrac{1}{4}0\tfrac{1}{4}/1\bar{1}\bar{1})^{1/3}_{+}, & (111)^{2/3}\mathscr{B}^{1/2}\mathscr{C}^{1/2}B \Rightarrow (\bar{\tfrac{1}{4}}\tfrac{1}{4}0/\bar{1}\bar{1}1)^{2/3}_{+}\\(111)^{1/3}\mathscr{C}^{1/2}\mathscr{A}^{1/2}C = (\tfrac{1}{4}\tfrac{1}{4}0/\bar{1}1\bar{1})^{1/3}_{+}, & (111)^{2/3}\mathscr{C}^{1/2}\mathscr{A}^{1/2}C \Rightarrow (0\tfrac{1}{4}\tfrac{1}{4}/1\bar{1}\bar{1})^{2/3}_{+}\end{array}\right\} \cdot$$
$$\tag{27}$$

Now

$$[(111)^{1/3}\mathscr{A}^{1/2}\mathscr{B}^{1/2}A]^2 = (\bar{\tfrac{1}{4}}\tfrac{1}{4}0/\bar{1}\bar{1}1)^{2/3}_{+}, \quad [(111)^{1/3}\mathscr{A}^{1/2}\mathscr{B}^{1/2}A]^3 = I$$

so demonstrating the existence of a cyclic group

$$\{\{(111)^{1/3}\mathscr{A}^{1/2}\mathscr{B}^{1/2}A\}\}_3 = \{I, (0\tfrac{1}{4}\tfrac{1}{4}/\bar{1}\bar{1}1)^{1/3}_{+}, (\bar{\tfrac{1}{4}}\tfrac{1}{4}0/\bar{1}\bar{1}1)^{2/3}_{+}\}$$

i.e. a realisation of $\{3\}$ within $\{2_13\}$, which describes a 3-fold rotation axis parallel to $[\bar{1}\bar{1}1]$. Here and in future work the symbol $\{\{\ \}\}_n$ signifies a cyclic group or groupoid of order n generated by the operator shown. Similarly there exist the further cyclic groups

$$\{\{(111)^{1/3}\mathscr{B}^{1/2}\mathscr{C}^{1/2}B\}\}_3, \quad \{\{(111)^{1/3}\mathscr{C}^{1/2}\mathscr{A}^{1/2}C\}\}_3$$

and also of course $\{\{(111)^{1/3}\}\}_3$, within $\{2_13\}$. Our analysis demonstrates that the 2-fold rotation axes of $\{23\}$ may be consistently replaced by 2-fold screw axes, whilst still preserving the 3-fold rotation axes (though translated from their original intersecting locations), so accounting for the symbolism $\{2_13\}$. The same type of analysis shows that no other groupoid could be constructed from $\{23\}$, e.g. the set

$$\{222\} + \mathscr{A}^{1/3}\mathscr{B}^{1/3}\mathscr{C}^{1/3}(111)^{1/3}\{2_12_12_1\} + \mathscr{A}^{2/3}\mathscr{B}^{2/3}\mathscr{C}^{2/5}\{2_12_12_1\} \quad (28)$$

would not be admissible since some of its operators are neither rotations nor screws, e.g. the product

$$\mathscr{A}^{1/3}\mathscr{B}^{1/3}\mathscr{C}^{1/3}(111)^{1/3} . \mathscr{A}^{1/2}\mathscr{B}^{1/2}(100)^{1/2} = \mathscr{A}^{1/3}\mathscr{B}^{1/3}\mathscr{C}^{1/3} . \mathscr{B}^{1/2}\mathscr{C}^{1/2}$$
$$(111)^{1/3} . (100)^{1/2}$$
$$= \mathscr{A}^{1/3}\mathscr{B}^{5/6}\mathscr{C}^{5/6}(\bar{1}\bar{1}1)^{1/3}$$

carries no geometrical significance.

Starting with the octahedral group

$$\{432\} = \{23\} + (100)^{1/4}\{23\} \quad\quad\quad\quad\quad\quad\quad\quad (29)$$

introduced in (7), Chapter 3, we construct the octahedral groupoid

$$\{4_132\} = \{2_13\} + \mathscr{A}^{1/4}\mathscr{B}^{1/4}\mathscr{C}^{-1/4}(100)^{1/4}\{2_13\}, \quad\quad\quad (30)$$

where

$$\mathscr{A}^{1/4}\mathscr{B}^{1/4}\mathscr{C}^{-1/4}(100)^{1/4} = [100]^{1/4}(0\tfrac{1}{8}\tfrac{\bar{1}}{8}/100)^{1/4}_{+}. \quad\quad (31)$$

This operator is seen to be a four-fold screw, with the factor $\mathscr{A}^{1/4}\mathscr{B}^{1/4}\mathscr{C}^{-1/4}$ (rather than just $\mathscr{A}^{1/4}$) necessary to ensure that

$$[\mathscr{A}^{1/4}\mathscr{B}^{1/4}\mathscr{C}^{-1/4}(100)^{1/4}]^2 = \mathscr{A}^{1/2}\mathscr{B}^{1/2}(100)^{1/2} = [100]^{1/2}(0\tfrac{1}{4}0/100)^{1/2} \subset \{2_13\}.$$
$$(32)$$

Following the methods of Appendix 1, we find (omitting possible factors $\mathscr{A}^{\pm1}$, $\mathscr{B}^{\pm1}$, $\mathscr{C}^{\pm1}$ on the right-hand side):

$$\mathscr{A}^{1/4}\mathscr{B}^{1/4}\mathscr{C}^{-1/4}(100)^{1/4} . I = [100]^{1/4}(0\tfrac{1}{8}\tfrac{\bar{1}}{8}/100)^{1/4}_{+}$$
$$\text{,,} \quad\quad\quad .\mathscr{A}^{1/2}\mathscr{B}^{1/2}A = [100]^{3/4}(0\tfrac{1}{8}\tfrac{1}{8}/100)^{3/4}_{+}$$
$$\text{,,} \quad\quad\quad .\mathscr{B}^{1/2}\mathscr{C}^{1/2}B = \quad\quad (\tfrac{1}{8}\tfrac{\bar{1}}{8}\tfrac{1}{8}/011)^{1/2}$$
$$\text{,,} \quad\quad\quad .\mathscr{C}^{1/2}.\mathscr{A}^{1/2}C = \quad\quad (\tfrac{3}{8}\tfrac{3}{8}\tfrac{3}{8}/01\bar{1})^{1/2}$$

$$\mathscr{A}^{1/4}\mathscr{B}^{1/4}\mathscr{C}^{-1/4}(100)^{1/4}(111)^{1/3}.I = (\tfrac{11\bar{1}}{888}/101)^{1/2}$$

$$,, \qquad\qquad .\mathscr{A}^{1/2}\mathscr{B}^{1/2}A = [010]^{3/4}(\tfrac{1}{8}0\tfrac{1}{8}/010)^{3/4}_+$$

$$,, \qquad\qquad .\mathscr{B}^{1/2}\mathscr{C}^{1/2}B = (\tfrac{333}{888}/10\bar{1})^{1/2}$$

$$,, \qquad\qquad .\mathscr{C}^{1/2}\mathscr{A}^{1/2}C = [010]^{1/4}(\tfrac{\bar{1}}{8}0\tfrac{1}{8}/010)^{1/4}_+$$

$$\mathscr{A}^{1/4}\mathscr{B}^{1/4}\mathscr{C}^{-1/4}(100)^{1/4}(111)^{2/3}.I = [001]^{3/4}(\tfrac{1}{8}\tfrac{1}{8}0/001)^{3/4}_+$$

$$,, \qquad\qquad .\mathscr{A}^{1/2}\mathscr{B}^{1/2}A = (\tfrac{333}{888}/1\bar{1}0)^{1/2}$$

$$,, \qquad\qquad .\mathscr{B}^{1/2}\mathscr{C}^{1/2}B = (\tfrac{\bar{1}11}{888}/110)^{1/2}$$

$$,, \qquad\qquad .\mathscr{C}^{1/2}\mathscr{A}^{1/2}C = [001]^{1/4}(\tfrac{1}{8}\tfrac{\bar{1}}{8}0/001)^{1/4}_+.$$

$$(33)$$

All these operators carry geometrical significance either as 2-fold rotations, or as 4-fold screws, and we note the property

$$[\mathscr{A}^{1/4}\mathscr{B}^{1/4}\mathscr{C}^{-1/4}(100)^{1/4}]^3 = [100]^{3/4}(0\tfrac{1}{8}\tfrac{1}{8}/100)^{3/4}_+, \qquad (34)$$

which, together with (32), demonstrates the existence of a cyclic groupoid $\{\{\mathscr{A}^{1/4}\mathscr{B}^{1/4}\mathscr{C}^{-1/4}(100)^{1/4}\}\}_4$ within $\{4_132\}$, which describes a 4-fold screw axis parallel to [100]. Similarly there exist the further cyclic groupoids

$$\{\{\mathscr{A}^{-1/4}\mathscr{B}^{1/4}\mathscr{C}^{1/4}(010)^{1/4}\}\}_4, \quad \{\{\mathscr{A}^{1/4}\mathscr{B}^{-1/4}\mathscr{C}^{1/4}(001)^{1/4}\}\}_4$$

and the cyclic groups

$$\{\{\mathscr{A}^{1/4}\mathscr{B}^{-1/4}\mathscr{C}^{1/4}(011)^{1/2}\}\}_2, \text{ etc.,}$$

within $\{4_132\}$. This analysis shows that the 4-fold rotation axes of $\{432\}$ may be consistently replaced by 4-fold screw axes, whilst still preserving the 3-fold and 2-fold rotation axes (though translated from their original intersecting locations), so accounting for the symbolism $\{4_132\}$. A similar analysis holds for

$$\{4_232\} = \{23\} + \mathscr{A}^{1/2}\mathscr{B}^{1/2}\mathscr{C}^{1/2}(100)^{1/2}\{23\},$$
$$\{4_332\} = \{2_13\} + \mathscr{A}^{3/4}\mathscr{B}^{3/4}\mathscr{C}^{-3/4}(100)^{3/4}\{2_13\} \qquad (35)$$

which share with $\{4_132\}$ the isomorphic equivalence

$$\{4_132\}, \{4_232\}, \{4_332\} \Leftrightarrow \{432\} \bmod \{\mathscr{T}\} \qquad (36)$$

Collecting all the groupoids of this chapter, we arrive at the non-Bravais space groups listed in Table 11.1.

Chapter 12

Glide Planes

12.1 GLIDE OPERATORS

We now introduce the symbol $(hkl)_m$ to signify a reflection through the lattice plane (hkl), e.g.

$$M = (001)_m, \quad D = (010)_m, \text{ in Chapter 2. Of course } (hkl)_m^2 = I.$$

Proceeding as for screws, $(hkl)_m$ may be generalised to the composite operator $\mathscr{A}^\lambda \mathscr{B}^\mu \mathscr{C}^\nu (hkl)_m$, which signifies a reflection followed by a translation $\lambda a + \mu b + \nu c$. If this lies parallel to the plane (hkl), we obtain a glide reflection with the property

$$\mathscr{A}^\lambda \mathscr{B}^\mu \mathscr{C}^\nu (hkl)_m = (hkl)_m \, \mathscr{A}^\lambda \mathscr{B}^\mu \mathscr{C}^\nu, \tag{1}$$

so yielding the powers

$$\mathscr{A}^{2\lambda} \mathscr{B}^{2\mu} \mathscr{C}^{2\nu}, \quad \mathscr{A}^{3\lambda} \mathscr{B}^{3\mu} \mathscr{C}^{3\nu} (hkl)_m, \text{ etc.} \tag{2}$$

Since $\mathscr{A}^{2\lambda} \mathscr{B}^{2\mu} \mathscr{C}^{2\nu}$ must be a lattice translation operator it follows that

$$\mathscr{A}^\lambda \mathscr{B}^\mu \mathscr{C}^\nu = \mathscr{A}^{\pm 1/2}, \quad \mathscr{A}^{\pm 1/2} \mathscr{B}^{\pm 1/2}, \quad \mathscr{A}^{\pm 1/2} \mathscr{B}^{\pm 1/2} \mathscr{C}^{\pm 1/2}, \text{ etc.,} \tag{3}$$

as exemplified by the common glide reflections

$$\mathscr{A}^{1/2}(001)_m, \quad \mathscr{B}^{1/2}(001)_m, \quad \mathscr{A}^{1/2} \mathscr{B}^{1/2}(001)_m, \text{ etc.} \tag{4}$$

$$\mathscr{A}^{1/2} \mathscr{B}^{-1/2}(110)_m, \quad \mathscr{C}^{1/2}(110)_m, \quad \mathscr{A}^{1/2} \mathscr{B}^{-1/2} \mathscr{C}^{1/2}(110)_m, \text{ etc.} \tag{5}$$

If $\lambda a + \mu b + \nu c$ lies normal to (hkl), there results a reflection through a parallel plane separated from the original by $\frac{1}{2}(\lambda a + \mu b + \nu c)$. This is indicated by writing

$$\mathscr{A}^\lambda \mathscr{B}^\mu \mathscr{C}^\nu (hkl)_m = \left(\frac{\lambda}{2} \frac{\mu}{2} \frac{\nu}{2} \middle/ hkl \right)_m, \tag{6}$$

where now

$$\mathscr{A}^\lambda \mathscr{B}^\mu \mathscr{C}^\nu (hkl)_m = (hkl)_m \mathscr{A}^{-\lambda} \mathscr{B}^{-\mu} \mathscr{C}^{-\nu} \tag{7}$$

by contrast with (1). Our analysis also holds if λ, μ, ν are integers, but in that case

$$\mathscr{A}^\lambda \mathscr{B}^\mu \mathscr{C}^\nu (hkl)_m \Rightarrow (hkl)_m \text{ in parallel with (3), Chapter 11.}$$

12.2 ABELIAN GLIDE GROUPOIDS

Corresponding with the point group

$$\{m\} = \{I, (001)_m\}, \tag{9}$$

we construct the groupoids

$$\{a\} = \{I, \mathscr{A}^{1/2}(001)_m\}; \quad \text{axis glide} \tag{10}$$

$$\{b\} = \{I, \mathscr{B}^{1/2}(001)_m\}; \quad ,, \quad ,, \tag{11}$$

$$\{g\} = \{I, \mathscr{A}^{1/2}\mathscr{B}^{1/2}(001)_m\}; \quad \text{diagonal glide.} \tag{12}$$

Clearly

$$\{a\}, \quad \{b\}, \quad \{g\} \Leftrightarrow \{m\} \bmod \{\mathscr{T}\}, \tag{13}$$

which implies that $\{a\} \otimes \{\mathscr{T}\}$, etc., are space groups providing the same macroscopic symmetry as does $\{m\} \otimes \{\mathscr{T}\}$, i.e. that characterised by the point-group $\{m\}$. No physical or geometrical distinction can be drawn between (10), (11) or (12), since [010], [110] could equally well be labelled [100]. Hence only (10) appears in Table 12.1.

Corresponding with the point group

$$\left\{\frac{2}{m}\right\} = \{2\} + (001)_m\{2\} = \{I, (001)^{1/2}, (001)_m, J\}, \tag{14}$$

we construct the groupoids

$$\left\{\frac{2}{a}\right\} = \{2\} + \mathscr{A}^{1/2}(001)_m\{2\} = \{I, (001)^{1/2}, \mathscr{A}^{1/2}(001)_m, \mathscr{A}^{1/2}J\} \tag{15}$$

$$\left\{\frac{2_1}{m}\right\} = \{2_1\} + (001)_m\{2_1\} = \{I, \mathscr{C}^{1/2}(001)^{1/2}, (001)_m, \mathscr{C}^{-1/2}J\} \tag{16}$$

$$\left\{\frac{2_1}{a}\right\} = \{2_1\} + \mathscr{A}^{1/2}(001)_m\{2_1\} = \{I, \mathscr{C}^{1/2}(001)^{1/2}, \mathscr{A}^{1/2}(001)_m, \mathscr{A}^{1/2}\mathscr{C}^{-1/2}J\} \tag{17}$$

bearing in mind

$$(001)_m(001)^{1/2} = (001)^{1/2}(001)_m = J, \tag{18}$$

$$(001)_m \, \mathscr{C}^\nu = \mathscr{C}^{-\nu}(001)_m. \tag{19}$$

All the operators of (14) have a clear geometrical significance. This also holds for the corresponding operators of (15)–(17), which are either 2-fold rotations or screws; pure or glide reflections; and inversions. The following theorem on inversions is relevant:

> if J signifies an inversion through the point $[0, 0, 0]$, then $\mathscr{T}J$ signifies an inversion through the point $\mathscr{T}^{1/2}[0, 0, 0]$. (20)

Equivalent representations for (14)–(17) are

$$\left\{\frac{2}{m}\right\} = \{2\} + J\{2\}, \quad \left\{\frac{2}{a}\right\} = \{2\} + \mathscr{A}^{1/2}J\{2\}$$

$$\left\{\frac{2_1}{m}\right\} = \{2_1\} + J\{2_1\}, \quad \left\{\frac{2_1}{a}\right\} = \{2_1\} + \mathscr{A}^{1/2}J\{2_1\} \tag{21}$$

on bearing in mind the counterparts of (18), (19):

$$J(hkl)^{1/2} = (hkl)^{1/2}J = (hkl)_m, \tag{22}$$

$$J\mathscr{T} = \mathscr{T}^{-1}J. \tag{23}$$

Compared with $(001)_m$, J has the two main advantages: it is isotropic and it commutes with every point-group operator. Hence it proves to be indispensable for the cubic point-groups and groupoids, but it would carry no advantage with $\left\{\frac{2}{m}\right\}$, $\left\{\frac{4}{m}\right\}$, $\left\{\frac{6}{m}\right\}$, $\{2mm\}$, $\{3m\}$, ... etc. Finally in line with the equivalence of (11), (12) to (10), nothing new emerges on replacing $\mathscr{A}^{1/2}$ by $\mathscr{B}^{1/2}$ or $\mathscr{A}^{1/2}\mathscr{B}^{1/2}$ in (15)–(17).

12.3 DIHEDRAL GLIDE GROUPOIDS

Starting with

$$\{2mm\} = \{2\} + (010)_m\{2\} = \{I, (001)^{1/2}, (010)_m, (100)_m\}, \tag{24}$$

where $(100)_m$, $(010)_m$ signify reflections through the planes $x = 0$, $y = 0$ respectively (Fig. 12.1), and where

$$(010)_m (001)^{1/2} = (001)^{1/2}(010)_m = (100)_m, \tag{25}$$

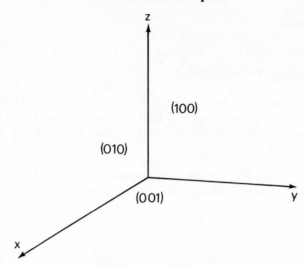

Fig. 12.1—Reflection planes (100), (010), (001), i.e. planes $x=0$, $y=0$, $z=0$ respectively. The operators $(100)_m$, $(010)_m$, $(001)_m$ signify reflections through these planes respectively.

we construct the groupoids

$$\{2am\} = \{2\} + \mathscr{A}^{1/2}(010)_m\{2\} = \{I, (001)^{1/2}, \mathscr{A}^{1/2}(010)_m, \mathscr{A}^{1/2}(100)_m\} \tag{26}$$

$$\{2cc\} = \{2\} + \mathscr{C}^{1/2}(010)_m\{2\} = \{I, (001)^{1/2}, \mathscr{C}^{1/2}(010)_m, \mathscr{C}^{1/2}(100)_m\} \tag{27}$$

$$\{2gc\} = \{2\} + \mathscr{C}^{1/2}\mathscr{A}^{1/2}(010)_m\{2\} = \{I, (001)^{1/2}, \mathscr{C}^{1/2}\mathscr{A}^{1/2}(010)_m,$$
$$\mathscr{C}^{1/2}\mathscr{A}^{1/2}(100)_m\} \tag{28}$$

$$\{2ab\} = \{2\} + \mathscr{A}^{1/2}\mathscr{B}^{1/2}(010)_m\{2\} = \{I, (001)^{1/2}, \mathscr{A}^{1/2}\mathscr{B}^{1/2}(010)_m,$$
$$\mathscr{A}^{1/2}\mathscr{B}^{1/2}(100)_m\} \tag{29}$$

$$\{2gg\} = \{2\} + \mathscr{A}^{1/2}\mathscr{B}^{1/2}\mathscr{C}^{1/2}(010)_m\{2\} = \{I, (001)^{1/2}, \mathscr{A}^{1/2}\mathscr{B}^{1/2}\mathscr{C}^{1/2}(010)_m,$$
$$\mathscr{A}^{1/2}\mathscr{B}^{1/2}\mathscr{C}^{1/2}(100)_m\}. \tag{30}$$

These are labelled in accordance with the glide reflections which appear in the third and fourth columns utilising the terminology of (10)–(12), e.g.

$$\mathscr{A}^{1/2}\mathscr{B}^{1/2}(010)_m = \mathscr{A}^{1/2}(0\tfrac{1}{4}0/010)_m \text{ i.e. an axis glide on (010)} \tag{31}$$

$$\mathscr{A}^{1/2}\mathscr{B}^{1/2}\mathscr{C}^{1/2}(100)_m = \mathscr{B}^{1/2}\mathscr{C}^{1/2}(\tfrac{1}{4}00/100)_m, \text{ i.e. a diagonal glide on}$$
$$(100). \tag{32}$$

By virtue of symmetry, nothing new emerges on replacing $\mathscr{A}^{1/2}$ by $\mathscr{B}^{1/2}$ in any of these, and it may be verified that the hypothetical groupoids

$$\{2mc\}, \quad \{2ac\}, \quad \{2mm\}, \quad \{2ag\} \tag{33}$$

can not be mathematically constructed.

Replacing 2 by 2_1 in $\{2mm\}$ produces the groupoid

$$\{2_1mc\} = \{2_1\} + (010)_m\{2_1\} = \{I, \mathscr{C}^{1/2}(001)^{1/2}, (010)_m, \mathscr{C}^{1/2}(100)_m\}, \tag{34}$$

showing incidentally that $\{2_1mm\}$ is not admissible. It will be seen that $\{2_1mc\}$ appears directly on replacing 2 by 2_1 in $\{2mc\}$, from the set (33). Similarly replacing 2 by 2_1 in $\{2ac\}$, etc., produces the new groupoids

$$\{2_1ac\} = \{2_1\} + \mathscr{A}^{1/2}(010)_m\{2_1\} = \{I, \mathscr{C}^{1/2}(001)^{1/2}, \mathscr{A}^{1/2}(010)_m, \\ \mathscr{A}^{1/2}\mathscr{C}^{1/2}(100)_m\} \tag{35}$$

$$\{2_1gm\} = \{2_1\} + \mathscr{C}^{1/2}\mathscr{A}^{1/2}(010)_m\{2_1\} = \{I, \mathscr{C}^{1/2}(001)^{1/2}, \mathscr{C}^{1/2}\mathscr{A}^{1/2}(010)_m, \\ \mathscr{A}^{1/2}(100)_m\} \tag{36}$$

$$\{2_1ag\} = \{2_1\} + \mathscr{A}^{1/2}\mathscr{B}^{1/2}(010)_m\{2_1\} = \{I, \mathscr{C}^{1/2}(001)^{1/2}, \mathscr{A}^{1/2}\mathscr{B}^{1/2}(010)_m, \\ \mathscr{A}^{1/2}\mathscr{B}^{1/2}\mathscr{C}^{1/2}(100)_m\} \tag{37}$$

though it is conventional to list $\{2_1gb\}$ in place of $\{2_1ag\}$. These complete the patterns based upon (24).

Starting with

$$\{4mm\} = \{4\} + (010)_m\{4\} = \begin{Bmatrix} I, (001)^{1/4}, (001)^{2/4}, (001)^{3/4} \\ (010)_m, (110)_m, (100)_m, (1\bar{1}0)_m \end{Bmatrix}, \tag{38}$$

where

$$(010)_m (001)^{1/4} = (110)_m, (010)_m (001)^{2/4} = (100)_m, (010)_m (001)^{3/4} = (1\bar{1}0)_m, \tag{39}$$

we construct the groupoids

$$\{4am\} = \{4\} + \mathscr{A}^{1/2}\mathscr{B}^{1/2}(010)_m\{4\} \tag{40}$$

$$\{4cc\} = \{4\} + \mathscr{C}^{1/2}(010)_m\{4\} \tag{41}$$

$$\{4mc\} = \{4\} + \mathscr{A}^{1/2}\mathscr{B}^{1/2}\mathscr{C}^{1/2}(010)_m\{4\} \tag{42}$$

on bearing in mind (31), (32), etc., and noting that

$$\mathscr{A}^{1/2}\mathscr{B}^{1/2}(110)_m = (\tfrac{1}{4}\tfrac{1}{4}0/110)_m, \quad \mathscr{A}^{1/2}\mathscr{B}^{1/2}(1\bar{1}0)_m = \mathscr{B} \cdot \mathscr{A}^{1/2}\mathscr{B}^{-1/2}(1\bar{1}0)_m \\ \Rightarrow (\tfrac{1}{4}\tfrac{\bar{1}}{4}0/1\bar{1}0)_m. \tag{43}$$

Certainly the operator $\mathscr{A}^{1/2}\mathscr{B}^{1/2}(1\bar{1}0)_m$ could be interpreted as a glide reflection, since the translation $\tfrac{1}{2}(\mathbf{a}+\mathbf{b})$ lies parallel to the plane $x - y = 0$, but it is preferably transformed into the pure reflection operator $\mathscr{A}^{1/2}\mathscr{B}^{-1/2}$ $(1\bar{1}0)_m$ by factors from $\{\mathscr{T}\}$. The only other possibilities based upon $\{4mm\}$

prove to be

$$\left.\begin{array}{l}\{4_2mc\}=\{4_2\}+(010)_m\{4_2\}, \{4_2cm\}=\{4_2\}+\mathscr{C}^{1/2}(010)_m\{4_2\}\\\{4_2ac\}=\{4_2\}+\mathscr{A}^{1/2}\mathscr{B}^{1/2}\{4_2\}, \{4_2gm\}=\{4_2\}+\mathscr{A}^{1/2}\mathscr{B}^{1/2}\mathscr{C}^{1/2}(010)_m\{4_2\}\end{array}\right\} \quad (44)$$

where $\{4_2mc\}$ provides the reflections $(010)_m$, $(001)_m$ whilst $\{4_2cm\}$ provides the reflections $(110)_m$, $(1\bar{1}0)_m$.

From

$$\{\bar{4}m2\}=\{\bar{4}\}+(010)_m\{\bar{4}\}$$
$$=\left\{\begin{array}{l}I, J(001)^{1/4}, (001)^{1/2}, J(001)^{3/4}\\(010)_m, (110)^{1/2}, (100)_m, (1\bar{1}0)^{1/2}\end{array}\right\}, \quad (45)$$

where

$$J(010)_m (001)^{1/4}=(110)^{1/2}, \quad J(010)_m (001)^{3/4}=(1\bar{1}0)^{1/2}, \quad (46)$$

we obtain

$$\left.\begin{array}{l}\{\bar{4}c2\}=\{\bar{4}\}+\mathscr{C}^{1/2}(010)_m\{\bar{4}\}, \{\bar{4}a2\}=\mathscr{A}^{1/2}\mathscr{B}^{1/2}(010)_m\{\bar{4}\}\\\{\bar{4}g2\}=\{\bar{4}\}+\mathscr{A}^{1/2}\mathscr{B}^{1/2}\mathscr{C}^{1/2}(010)_m\{\bar{4}\}\end{array}\right\} \quad (47)$$

bearing in mind

$$\left.\begin{array}{l}\mathscr{C}^{1/2}(110)^{1/2}=(00\tfrac{1}{4}/110)^{1/2}, \mathscr{A}^{1/2}\mathscr{B}^{1/2}(1\bar{1}0)^{1/2}=(\tfrac{1}{4}\tfrac{\bar{1}}{4}0/1\bar{1}0)^{1/2}\\\mathscr{A}^{1/2}\mathscr{B}^{1/2}(110)^{1/2}=\mathscr{B}.\mathscr{A}^{1/2}\mathscr{B}^{-1/2}(110)^{1/2}\Rightarrow(\tfrac{1}{4}\tfrac{\bar{1}}{4}0/110)^{1/2}\end{array}\right\}. \quad (48)$$

Similarly, from

$$\{\bar{4}2m\}=\{\bar{4}\}+(100)^{1/2}\{\bar{4}\}$$
$$=\left\{\begin{array}{l}I J(001)^{1/4}, (001)^{1/2}, J(001)^{3/4}\\(100)^{1/2}, (1\bar{1}0)_m, (010)^{1/2}, (110)_m\end{array}\right\}, \quad (49)$$

we obtain

$$\left.\begin{array}{l}\{\bar{4}2c\}=\{\bar{4}\}+\mathscr{C}^{1/2}(100)^{1/2}\{\bar{4}\}, \{\bar{4}2_1m\}=\{\bar{4}\}+\mathscr{A}^{1/2}\mathscr{B}^{1/2}(100)^{1/2}\{\bar{4}\}\\\{\bar{4}2_1c\}=\{\bar{4}\}+\mathscr{A}^{1/2}\mathscr{B}^{1/2}\mathscr{C}^{1/2}(100)^{1/2}\{\bar{4}\}\end{array}\right\}, \quad (50)$$

bearing in mind

$$\mathscr{A}^{1/2}\mathscr{B}^{1/2}\mathscr{C}^{1/2}(100)^{1/2}=[100]^{1/2}(0\tfrac{1}{4}\tfrac{1}{4}/100)^{1/2}, \text{ etc.} \quad (51)$$

$$\mathscr{A}^{1/2}\mathscr{B}^{1/2}\mathscr{C}^{1/2}(110)_m=\mathscr{C}^{1/2}(\tfrac{1}{4}\tfrac{1}{4}0/110)_m, \text{ i.e. an axis glide on } (110)_m$$
etc. $\quad (52)$

From

$$\{6mm\} = \{6\} + (010)_m\{6\}, \tag{53}$$

we obtain without difficulty

$$\left. \begin{array}{l} \{6cc\} = \{6\} + \mathscr{C}^{1/2}\,(010)_m\{6\} \\ \{6_3mc\} = \{6_3\} + (010)_m\{6_3\},\ \{6_3cm\} = \{6_3\} + \mathscr{C}^{1/2}\,(010)_m\{6_3\} \end{array} \right\}. \tag{54}$$

Finally, from

$$\{3m\} = \{3m1\},\ \{31m\} = \{3\} + (010)_m\{3\}, \tag{55}$$

we obtain

$$\{3c1\},\ \{31c\} = \{3\} + \mathscr{C}^{1/2}\,(010)_m\{3\} \tag{56}$$

utilising the alternative settings for the secondary axes displayed in Fig. 9.3.

The results of this section may be summarised as follows:

$$\left. \begin{array}{l} \{2am\},\ \{2cc\},\ \{2gc\},\ \{2ab\},\ \{2ag\} \\ \{2_1mc\},\ \{2_1ac\},\ \{2_1gm\},\ \{2_1ag\} \end{array} \right\} \Leftrightarrow \{2mm\}\ \text{mod}\ \{\mathscr{T}\} \tag{57}$$

$$\left. \begin{array}{l} \{4am\},\ \{4cc\},\ \{4gc\} \\ \{4_2mc\},\ \{4_2cm\},\ \{4_2ac\},\ \{4_2gm\} \end{array} \right\} \Leftrightarrow \{4mm\}\ \text{mod}\ \{\mathscr{T}\} \tag{58}$$

$$\{\bar{4}c2\},\ \{\bar{4}a2\},\ \{\bar{4}g2\} \qquad\qquad \Leftrightarrow \{\bar{4}m2\}\ \text{mod}\ \{\mathscr{T}\} \tag{59}$$

$$\{\bar{4}2c\},\ \{\bar{4}2_1m\},\ \{\bar{4}\,2_1c\} \qquad\qquad \Leftrightarrow \{\bar{4}2m\}\ \text{mod}\ \{\mathscr{T}\} \tag{60}$$

$$\{6cc\},\ \{6_3mc\},\ \{6_3cm\} \qquad\qquad \Leftrightarrow \{6mm\}\ \text{mod}\ \{\mathscr{T}\} \tag{61}$$

$$\{3c1\},\ \{31c\} \qquad\qquad\qquad \Leftrightarrow \{3m1\},\ \{31m\}\ \text{mod}\ \{\mathscr{T}\} \tag{62}$$

so providing the relevant space groups listed in Table 12.1.

12.4 FURTHER ORTHORHOMBIC GROUPOIDS

The groupoids (57) above all appertain to the orthorhombic system. Further orthorhombic groupoids may be constructed starting from

$$\left\{ \frac{2}{m}\,\frac{2}{m}\,\frac{2}{m} \right\} = \{222\} + J\{222\} \tag{63}$$

$$= \left\{ \begin{array}{l} I,\ (001)^{1/2},\ (100)^{1/2},\ (010)^{1/2} \\ J,\ (001)_m,\ (100)_m,\ (010)_m \end{array} \right\},$$

which immediately provides

$$\left\{\frac{2}{m}\frac{2}{c}\frac{2}{c}\right\}=\{222\}+\mathscr{C}^{1/2}J\{222\}, \quad \left\{\frac{2}{g}\frac{2}{b}\frac{2}{a}\right\}=\{222\}+\mathscr{A}^{1/2}\mathscr{B}^{1/2}J\{222\}$$

$$\left\{\frac{2}{g}\frac{2}{g}\frac{2}{g}\right\}=\{222\}+\mathscr{A}^{1/2}\mathscr{B}^{1/2}\mathscr{C}^{1/2}J\{222\}. \tag{64}$$

A screw variant of (63) is

$$\left\{\frac{2_1}{m}\frac{2}{m}\frac{2}{c}\right\}=\{2_122\}+J\{2_122\} \tag{65}$$

$$=\left\{\begin{matrix} I, & \mathscr{C}^{1/2}(001)^{1/2}, & (100)^{1/2}, & \mathscr{C}^{-1/2}(010)^{1/2} \\ J, & \mathscr{C}^{-1/2}(001)_m, & (100)_m, & \mathscr{C}^{1/2}(010)_m \end{matrix}\right\},$$

which provides

$$\left\{\frac{2_1}{a}\frac{2}{m}\frac{2}{g}\right\}=\{2_122\}+\mathscr{A}^{1/2}J\{2_122\}, \quad \left\{\frac{2_1}{g}\frac{2}{c}\frac{2}{a}\right\}=\{2_122\}+\mathscr{C}^{1/2}\mathscr{A}^{1/2}J\{2_122\}$$

$$\left\{\frac{2_1}{g}\frac{2}{g}\frac{2}{a}\right\}=\{2_122\}+\mathscr{A}^{1/2}\mathscr{B}^{1/2}\mathscr{C}^{1/2}J\{2_122\}. \tag{66}$$

A second screw variant of (63) is

$$\left\{\frac{2}{m}\frac{2_1}{b}\frac{2_1}{a}\right\}=\{22_12_1\}+J\{22_12_1\}$$

$$=\left\{\begin{matrix} I, & (001)^{1/2}, & \mathscr{A}^{1/2}\mathscr{B}^{1/2}(100)^{1/2}, & \mathscr{A}^{1/2}\mathscr{B}^{1/2}(010)^{1/2} \\ J, & (001)_m, & \mathscr{A}^{-1/2}\mathscr{B}^{-1/2}(100)_m, & \mathscr{A}^{-1/2}\mathscr{B}^{-1/2}(010)_m \end{matrix}\right\}, \tag{67}$$

which provides

$$\left\{\frac{2}{a}\frac{2_1}{b}\frac{2_1}{g}\right\}=\{22_12_1\}+\mathscr{A}^{1/2}J\{22_12_1\}, \quad \left\{\frac{2}{g}\frac{2_1}{m}\frac{2_1}{m}\right\}=\{22_12_1\}+\mathscr{A}^{1/2}\mathscr{B}^{1/2}J\{22_12_1\}$$

$$\left\{\frac{2}{m}\frac{2_1}{g}\frac{2_1}{g}\right\}=\{22_12_1\}+\mathscr{C}^{1/2}J\{22_12_1\}, \quad \left\{\frac{2}{a}\frac{2_1}{g}\frac{2_1}{c}\right\}=\{22_12_1\}+\mathscr{C}^{1/2}\mathscr{A}^{1/2}J\{22_12_1\}$$

$$\left\{\frac{2}{g}\frac{2_1}{c}\frac{2_1}{c}\right\}=\{22_12_1\}+\mathscr{A}^{1/2}\mathscr{B}^{1/2}\,\mathscr{C}^{1/2}J\{22_12_1\}. \tag{68}$$

Finally, we have the pair

$$\left\{\frac{2_1}{b}\frac{2_1}{c}\frac{2_1}{a}\right\}=\{2_12_12_1\}+J\{2_12_12_1\}, \quad \left\{\frac{2_1}{g}\frac{2_1}{m}\frac{2_1}{a}\right\}=\{2_12_12_1\}+\mathscr{C}^{1/2}J\{2_12_12_1\}. \tag{69}$$

All the groupoids listed in (64)–(69) have the property

$$\Leftrightarrow \left\{ \frac{2}{m} \frac{2}{m} \frac{2}{m} \right\} \bmod \{\mathcal{T}\}. \tag{70}$$

No other independent orthorhombic groupoids can be constructed using the glide reflections of this chapter. However, two new ones may be constructed using the diamond glide reflections of Chapter 13.

12.5 FURTHER TETRAGONAL GROUPOIDS

Starting from

$$\left\{ \frac{4}{m} \frac{2}{m} \frac{2}{m} \right\} = \{422\} + J\{422\} = \{422\} + M\{422\}; \quad M = (001)_m, \tag{71}$$

introduced in (41), Chapter 2, we construct the variants

$$\left\{ \frac{4}{m} \frac{2}{c} \frac{2}{c} \right\} = \{422\} + \mathscr{C}^{1/2}J\{422\}, \quad \left\{ \frac{4}{g} \frac{2}{b} \frac{2}{m} \right\} = \{422\} + \mathscr{A}^{1/2}\mathscr{B}^{1/2}J\{422\}$$

$$\left\{ \frac{4}{g} \frac{2}{g} \frac{2}{c} \right\} = \{422\} + \mathscr{A}^{1/2}\mathscr{B}^{1/2}\mathscr{C}^{1/2}J\{422\}. \tag{72}$$

From $\{42_12\}$—see (18), Chapter 11—we obtain

$$\left\{ \frac{42_12}{mbm} \right\} = \{42_12\} + J\{42_12\}, \quad \left\{ \frac{4}{m} \frac{2_1}{g} \frac{2}{c} \right\} = \{42_12\} + \mathscr{C}^{1/2}J\{42_12\} \tag{73}$$

$$\left\{ \frac{42_12}{gmm} \right\} = \{42_12\} + \mathscr{A}^{1/2}\mathscr{B}^{1/2}J\{42_12\}, \quad \left\{ \frac{4}{g} \frac{2_1}{c} \frac{2}{c} \right\} = \{42_12\} +$$

$$\mathscr{A}^{1/2}\mathscr{B}^{1/2}\mathscr{C}^{1/2}J\{42_12\}.$$

From $\{4_222\}$—see Chapter 11—we obtain

$$\left\{ \frac{4_2}{m} \frac{2}{m} \frac{2}{c} \right\} = \{4_222\} + J\{4_222\}, \quad \left\{ \frac{4_2}{m} \frac{2}{c} \frac{2}{m} \right\} = \{4_222\} + \mathscr{C}^{1/2}J\{4_222\} \tag{74}$$

$$\left\{ \frac{4_2}{g} \frac{2}{b} \frac{2}{c} \right\} = \{4_222\} + \mathscr{A}^{1/2}\mathscr{B}^{1/2}J\{4_222\}, \quad \left\{ \frac{4_2}{g} \frac{2}{g} \frac{2}{m} \right\} = \{4_222\} +$$

$$\mathscr{A}^{1/2}\mathscr{B}^{1/2}\mathscr{C}^{1/2}J\{4_222\}.$$

From $\{4_22_12\}$—see Chapter 11—we obtain

$$\left\{ \frac{4_2}{m} \frac{2_1}{b} \frac{2}{c} \right\} = \{4_22_12\} + J\{4_22_12\}, \quad \left\{ \frac{4_2}{m} \frac{2_1}{g} \frac{2}{m} \right\} = \{4_22_12\} + \mathscr{C}^{1/2}J\{4_22_12\} \tag{75}$$

$$\left\{\frac{4_2}{g}\frac{2_1}{m}\frac{2}{c}\right\}=\{4_22_12\}+\mathscr{A}^{1/2}\mathscr{B}^{1/2}J\{4_22_12\},\quad\left\{\frac{4_2}{g}\frac{2_1}{c}\frac{2}{m}\right\}=\{4_22_12\}+$$

$$\mathscr{A}^{1/2}\mathscr{B}^{1/2}\mathscr{C}^{1/2}J\{4_22_12\}.$$

All the groupoids listed in (71)–(75) have the property

$$\Leftrightarrow\left\{\frac{4}{m}\frac{2}{m}\frac{2}{m}\right\}\bmod\{\mathscr{T}\}. \tag{76}$$

It will be shown in Chapter 13 that one further tetragonal groupoid can be constructed using the glide reflections of this chapter. Also, five new ones may be constructed using the diamond glide reflections of Chapter 13.

12.6 FURTHER HEXAGONAL GROUPOIDS

Utilising the representation

$$\left\{\frac{3}{m}m2\right\}=\{321\}+M\{321\};\quad M=(001)_m, \tag{77}$$

where $M(100)^{1/2}=(010)_m$, etc., (Fig. 9.3), we readily construct the variant

$$\left\{\frac{3}{m}c2\right\}=\{321\}+\mathscr{C}^{1/2}M\{321\}\Leftrightarrow\left\{\frac{3}{m}m2\right\}\bmod\{\mathscr{T}\}. \tag{78}$$

Similarly, from

$$\left\{\bar{3}\frac{2}{m}1\right\}=\{321\}+J\{321\} \tag{79}$$

where $J(100)^{1/2}=(100)_m^{\perp}$, i.e. a plane perpendicular to [100], etc., we obtain

$$\left\{\bar{3}\frac{2}{c}1\right\}=\{321\}+\mathscr{C}^{1/2}J\{321\}\Leftrightarrow\left\{\bar{3}\frac{2}{m}1\right\}\bmod\{\mathscr{T}\}. \tag{80}$$

A completely parallel analysis, on the alternative setting for the secondary axes, yields

$$\left\{\frac{3}{m}2m\right\}=\{312\}+M\{312\},\quad\left\{\frac{3}{m}2c\right\}=\{312\}+\mathscr{C}^{1/2}M\{312\}\Leftrightarrow\left\{\frac{3}{m}2m\right\} \tag{81}$$

$$\left\{\bar{3}1\frac{2}{m}\right\}=\{312\}+J\{312\},\quad\left\{\bar{3}1\frac{2}{c}\right\}=\{312\}+\mathscr{C}^{1/2}J\{312\}\Leftrightarrow\left\{\bar{3}1\frac{2}{m}\right\}. \tag{82}$$

Starting from

$$\left\{\frac{6}{m}\frac{2}{m}\frac{2}{m}\right\} = \{622\} + J\{622\} = \{622\} + M\{622\}, \tag{83}$$

we construct the variant

$$\left\{\frac{6}{m}\frac{2}{c}\frac{2}{c}\right\} = \{622\} + \mathscr{C}^{1/2}J\{622\} \Leftrightarrow \left\{\frac{6}{m}\frac{2}{m}\frac{2}{m}\right\} \bmod \{\mathscr{T}\}, \tag{84}$$

and finally, from $\{6_322\}$, we obtain

$$\left\{\frac{6_3}{m}\frac{2}{m}\frac{2}{c}\right\} = \{6_322\} + J\{6_322\}, \left\{\frac{6_3}{m}\frac{2}{c}\frac{2}{m}\right\} = \{6_322\} + \mathscr{C}^{1/2}J\{6_322\} \Leftrightarrow$$

$$\left\{\frac{6}{m}\frac{2}{m}\frac{2}{m}\right\} \bmod \{\mathscr{T}\}. \tag{85}$$

No further independent hexagonal groupoids can be constructed.

12.7 CUBIC GROUPOIDS

Starting from

$$\left\{\frac{2}{m}\bar{3}\right\} = \{23\} + J\{23\}, \tag{86}$$

introduced in (8), Chapter 3, we construct the variant

$$\left\{\frac{2}{g}\bar{3}\right\} = \{23\} + \mathscr{A}^{1/2}\mathscr{B}^{1/2}\mathscr{C}^{1/2}J\{23\} \Leftrightarrow \left\{\frac{2}{m}\bar{3}\right\} \bmod \{\mathscr{T}\}, \tag{87}$$

which may be understood by noting that it includes subgroupoids of the form:

(1) $\{I, (100)^{1/2}\} + \mathscr{A}^{1/2}\mathscr{B}^{1/2}\mathscr{C}^{1/2}J\{I, (100)^{1/2}\},$ \hfill (88)

i.e. $\{2\} + \mathscr{A}^{1/2}\mathscr{B}^{1/2}\mathscr{C}^{1/2}(100)_m\{2\}$, i.e. $\left\{\frac{2}{g}\right\}$, on bearing in mind $(100)_m = J(100)^{1/2}$ and that $\mathscr{A}^{1/2}\mathscr{B}^{1/2}\mathscr{C}^{1/2}(100)_m = \mathscr{B}^{1/2}\mathscr{C}^{1/2}(\frac{1}{4}00/100)_m$;

(2) $\{I, (111)^{1/3}, (111)^{2/3}\} + \mathscr{A}^{1/2}\mathscr{B}^{1/2}\mathscr{C}^{1/2}J\{I,(111)^{1/3}, (111)^{2/3}\}$ \hfill (89)

i.e. $\{\{\mathscr{A}^{1/2}\mathscr{B}^{1/2}\mathscr{C}^{1/2}J(111)^{1/3}\}\}_6$, i.e. a realisation of $\{\bar{3}\}$, utilising the symbolism introduced in section 11.4, where $\{\{\ \}\}_n$ signifies a cyclic groupoid of order n generated by the operator shown.

From $\{2_1 3\}$—see (24), Chapter 11—we obtain

$$\left\{\frac{2_1}{a}\,\bar{3}\right\} = \{2_1 3\} + J\{2_1 3\} \Leftrightarrow \left\{\frac{2}{m}\,\bar{3}\right\} \bmod \{\mathscr{T}\}, \tag{90}$$

as follows by noting that it includes subgroupoids of the form:

(1) $\{I, \mathscr{A}^{1/2}\mathscr{B}^{1/2}(100)^{1/2}\} + J\{I, \mathscr{A}^{1/2}\mathscr{B}^{1/2}(100)^{1/2}\},$ (91)

i.e. $\{2_1\} + \mathscr{A}^{1/2}\mathscr{B}^{1/2}(100)_m\{2_1\}$, i.e. $\left\{\dfrac{2_1}{a}\right\}$,

on bearing in mind $\mathscr{A}^{-1/2}\,\mathscr{B}^{+1/2} \Rightarrow \mathscr{A}^{1/2}\,\mathscr{B}^{1/2}$;

(2) $\{I, \quad \mathscr{B}^{1/2}\mathscr{C}^{1/2}(\bar{1}\bar{1}1)^{1/3}, \quad \mathscr{A}^{1/2}\mathscr{B}^{1/2}(\bar{1}\bar{1}1)^{2/3}\} + J\{I, \quad \mathscr{B}^{1/2}\mathscr{C}^{1/2}(\bar{1}\bar{1}1)^{1/3},$
$\mathscr{A}^{1/2}\mathscr{B}^{1/2}(\bar{1}\bar{1}1)^{2/3}\},$ (92)

i.e. $\{\{\mathscr{B}^{1/2}\mathscr{C}^{1/2}J(\bar{1}\bar{1}1)^{1/3}\}\}_6$, etc., so providing realisations of $\{\bar{3}\}$.

Starting from

$$\{\bar{4}3m\} = \{23\} + J(100)^{1/4}\{23\}, \tag{93}$$

introduced in (16), Chapter 3, we construct the variant

$$\{\bar{4}3g\} = \{23\} + \mathscr{A}^{1/2}\mathscr{B}^{1/2}\mathscr{C}^{1/2}J(100)^{1/4}\{23\} \Leftrightarrow \{\bar{4}3m\} \bmod \{\mathscr{T}\}, \tag{94}$$

which includes subgroupoids of the form:

(1) $\{I, \mathscr{A}^{1/2}\mathscr{B}^{1/2}\mathscr{C}^{1/2}J(1\bar{1}0)^{1/2}\}$, i.e. $\{I, \mathscr{A}^{1/2}\mathscr{B}^{1/2}\mathscr{C}^{1/2}(1\bar{1}0)_m\}$, (95)

i.e. $\{g\}$ since $\mathscr{A}^{1/2}\mathscr{B}^{1/2}\mathscr{C}^{1/2}(1\bar{1}0)_m$ signifies a diagonal glide on the lattice plane $(1\bar{1}0)$;

(2) $\{I, \mathscr{A}^{1/2}\mathscr{B}^{1/2}\mathscr{C}^{1/2}J(100)^{1/4}, (100)^{1/2}, \mathscr{A}^{1/2}\mathscr{B}^{1/2}\mathscr{C}^{1/2}J(100)^{3/4}\},$
 (96)

i.e. $\{\{\mathscr{A}^{1/2}\mathscr{B}^{1/2}\mathscr{C}^{1/2}J(100)^{1/4}\}\}_4$ so providing a realisation of $\{\bar{4}\}$.

From the cubic group

$$\left\{\frac{4}{m}\,\bar{3}\,\frac{2}{m}\right\} = \{432\} + J\{432\}, \tag{97}$$

introduced in (9), Chapter 3, we construct the variant

$$\left\{\frac{4}{g}\bar{3}\frac{2}{g}\right\} = \{432\} + \mathscr{A}^{1/2}\mathscr{B}^{1/2}\mathscr{C}^{1/2}J\{432\} \Leftrightarrow \left\{\frac{4}{m}\,\bar{3}\,\frac{2}{m}\right\} \bmod \{\mathscr{T}\}, \tag{98}$$

which may be understood by an analysis extending that of (90), e.g. it includes subgroupoids of the form

$$\{\{(100)^{1/4}\}\}_4 + \mathscr{A}^{1/2}\mathscr{B}^{1/2}\mathscr{C}^{1/2}(100)_m\{\{(100)^{1/4}\}\}_4, \text{ i.e. } \left\{\frac{4}{g}\right\}. \tag{99}$$

Finally, from

$$\{4_2 32\} = \{23\} + \mathscr{A}^{1/2}\mathscr{B}^{1/2}\mathscr{C}^{1/2}(100)^{1/4}\{23\}, \tag{100}$$

we obtain without difficulty

$$\left\{\frac{4_2}{m}\bar{3}\frac{2}{g}\right\} = \{4_2 32\} + J\{4_2 32\} \Leftrightarrow \left\{\frac{4}{m}\bar{3}\frac{2}{m}\right\} \bmod \{\mathscr{T}\}, \tag{101}$$

$$\left\{\frac{4_2}{g}\bar{3}\frac{2}{m}\right\} = \{4_2 32\} + \mathscr{A}^{1/2}\mathscr{B}^{1/2}\mathscr{C}^{1/2}J\{4_2 32\} \Leftrightarrow \left\{\frac{4}{m}\bar{3}\frac{2}{m}\right\} \bmod \{\mathscr{T}\}. \tag{102}$$

Five new independent cubic groupoids may be constructed using the diamond glide reflections of Chapter 13. The results of this chapter are collected in Table 12.1 below.

Table 12.1—Glide Groupoids*

Crystal system	Lattice type	Groupoid
monoclinic	P	a
		$\dfrac{2}{a}, \dfrac{2_1}{a}, \dfrac{2_1}{m}$
orthorhombic	P	$2cc, \ 2gg, \ 2ab, \ 2gc, \ 2am$
		$2_1 mc, \ 2_1 gb, \ 2_1 gm, \ 2_1 ac$
		$\dfrac{2\,2\,2}{g\,g\,g}, \ \dfrac{2\,2\,2}{m\,c\,c}, \ \dfrac{2\,2\,2}{g\,b\,a}$
		$\dfrac{2_1\,2\,2}{m\,m\,c}, \ \dfrac{2_1\,2\,2}{g\,g\,a}, \ \dfrac{2_1\,2\,2}{a\,m\,g}, \ \dfrac{2_1\,2\,2}{a\,c\,a}$
		$\dfrac{2\,2_1\,2_1}{m\,b\,a}, \ \dfrac{2\,2_1\,2_1}{g\,c\,c}, \ \dfrac{2\,2_1\,2_1}{m\,g\,g}, \ \dfrac{2\,2_1\,2_1}{g\,m\,m}$
		$\dfrac{2\,2_1\,2_1}{a\,b\,m}, \ \dfrac{2\,2_1\,2_1}{a\,g\,c}, \ \dfrac{2_1\,2_1\,2_1}{b\,c\,a}, \ \dfrac{2_1\,2_1\,2_1}{g\,m\,a}$
tetragonal	P	$\dfrac{4}{g}, \ \dfrac{4_2}{m}, \ \dfrac{4_2}{g}$
		$4am, \ 4cc, \ 4gc$
		$4_2 mc, \ 4_2 bc, \ 4_2 cm, \ 4_2 gm$
		$\dfrac{4\,2\,2}{m\,c\,c}, \ \dfrac{4\,2\,2}{g\,b\,m}, \ \dfrac{4\,2\,2}{g\,g\,c}$

Crystal system	Lattice type	Groupoid
		$\frac{4}{m}\frac{2_1}{b}\frac{2}{m}$, $\frac{4}{m}\frac{2_1}{g}\frac{2}{c}$, $\frac{4}{g}\frac{2_1}{m}\frac{2}{m}$, $\frac{4}{g}\frac{2_1}{c}\frac{2}{c}$
		$\frac{4_2}{m}\frac{2}{m}\frac{2}{c}$, $\frac{4_2}{m}\frac{2}{c}\frac{2}{m}$, $\frac{4_2}{g}\frac{2}{b}\frac{2}{c}$, $\frac{4_2}{g}\frac{2}{g}\frac{2}{m}$
		$\frac{4_2}{m}\frac{2_1}{b}\frac{2}{c}$, $\frac{4_2}{m}\frac{2_1}{g}\frac{2}{m}$, $\frac{4_2}{g}\frac{2_1}{m}\frac{2}{c}$, $\frac{4_2}{g}\frac{2_1}{c}\frac{2}{m}$
		$\bar{4}2c$, $\bar{4}2_1m$, $\bar{4}2_1c$
		$\bar{4}c2$, $\bar{4}a2$, $\bar{4}g2$
hexagonal	P	$\frac{6_3}{m}$, $6cc$, 6_3mc, 6_3cm
		$\frac{6}{m}\frac{2}{c}\frac{2}{c}$, $\frac{6_3}{m}\frac{2}{m}\frac{2}{c}$, $\frac{6_3}{m}\frac{2}{c}\frac{2}{m}$
		$3c1$, $31c$
		$\frac{3}{m}c2$, $\frac{3}{m}2c$, $\bar{3}\frac{2}{c}1$, $\bar{3}1\frac{2}{c}$
cubic	P	$\frac{2}{g}\bar{3}$, $\frac{2_1}{a}\bar{3}$, $\bar{4}3g$
		$\frac{4_2}{m}\bar{3}\frac{2}{g}$, $\frac{4_2}{g}\bar{3}\frac{2}{m}$, $\frac{4}{g}\bar{3}\frac{2}{g}$

*The symbol g signifies diagonal glide in place of the conventional symbol v.

Chapter 13

Diamond Glide

13.1 SPACE GROUPS: CENTRED LATTICES

The theory of Chapters 10–12 refers to primitive space groups whose existence is formally ensured by the following theorem:

> *Given a groupoid $\{\Gamma\}$ with the property $\{\Gamma\} \Leftrightarrow \{G\} \bmod \{\mathcal{T}\}$, then $\{\Gamma\} \otimes \{\mathcal{T}\}$ is a space group if $\{G\} \otimes \{\mathcal{T}\}$ is a space group.* (1)

A particular case of this theorem has already appeared in (11), Chapter 10. More generally we write

$$\Gamma_i = \mathcal{A}^\lambda \mathcal{B}^\mu \mathcal{C}^\nu G_i; \quad \left. \begin{array}{l} G_i \subset \{G\}, \{\Gamma_i\} \subset \{\Gamma\} \\ 0 \leqslant |\lambda|, |\mu|, |\nu| < 1 \end{array} \right\} \tag{2}$$

and note that

$$\Gamma_i \mathcal{T}_1 \Gamma_i^{-1} = \mathcal{A}^\lambda \mathcal{B}^\mu \mathcal{C}^\nu G_i . \overline{\mathcal{T}_1} . G_i^{-1} \mathcal{C}^{-\nu} \mathcal{B}^{-\mu} \mathcal{A}^{-\lambda} = \mathcal{T}_2; \ \mathcal{T}_1, \mathcal{T}_2 \subset \{\mathcal{T}\} \tag{3}$$

on bearing in mind that $\{G\}$ satisfies the coupling relation (11), Chapter 8. The existence of $\{\Gamma\} \otimes \{\mathcal{T}\}$ then immediately follows from (3).

The theory can be extended to centred space groups by virtue of the following theorem:

> *Given a groupoid $\{\Gamma\}$ with the property $\{\Gamma\} \Leftrightarrow \{G\} \bmod \{\mathcal{T}\}$, then $\{\Gamma\} \otimes \{\mathcal{I}\}$, $\{\Gamma\} \otimes \{\mathcal{F}\}$, $\{\Gamma\} \otimes \{\mathcal{E}\}$, $\{\Gamma\} \otimes \{\mathcal{R}\}$ are space groups if, respectively, $\{G\} \otimes \{\mathcal{T}\}$, etc., are space groups.* (4)

This theorem is readily proved by an easy extension of the above argument. Thus, for instance, we note from (2) that

$$\Gamma_i \mathcal{A}^{1/2} \mathcal{B}^{1/2} \mathcal{C}^{1/2} \Gamma_i^{-1} = \mathcal{A}^{\pm 1/2} \mathcal{B}^{\pm 1/2} \mathcal{C}^{\pm 1/2}$$

$$\text{if} \quad G_i \mathcal{A}^{1/2} \mathcal{B}^{1/2} \mathcal{C}^{1/2} G_i^{-1} = \mathcal{A}^{\pm 1/2} \mathcal{B}^{\pm 1/2} \mathcal{C}^{\pm 1/2}. \tag{5}$$

Building upon Tables 11.1, 12.1, no difficulty arises in systematically constructing centred space groups appropriate to each crystal system, as was done for the Bravais space groups (see Table 9.1). However, two limitations arise in practice.

(1) Most of the resulting space groups prove to be redundant, either because they are essentially covered by existing space groups or because they are equivalent to cognate space groups sharing the same isomorphic property, e.g.

$$\{4_2\}\otimes\{\mathscr{I}\}=\{4\}\otimes\{\mathscr{I}\}, \quad \{4_3\}\otimes\{\mathscr{I}\}=\{4_1\}\otimes\{\mathscr{I}\}. \tag{6}$$

(2) This systematic procedure does not account for special space groups which fall outside the scope of Theorem (4), e.g. $\left\{\dfrac{4_1}{a}\right\}\otimes\{\mathscr{I}\}$ exists even though $\left\{\dfrac{4_1}{a}\right\}\otimes\{\mathscr{T}\}$ does not. Alternatively expressed, the body-centred tetragonal lattice admits a microscopic symmetry pattern denied to the primitive tetragonal lattice. Similarly, the introduction of diamond glide enables twelve space groups to be constructed which could not be associated with any primitive space lattice.

13.2 EQUIVALENT SPACE GROUPS

To prove that $\{2_1\}\otimes\{\mathscr{I}\}=\{2\}\otimes\{\mathscr{I}\}$, we display these as

$$\{2\}\otimes\{\mathscr{I}\}=\{I, C, \mathscr{A}^{1/2}\mathscr{B}^{1/2}\mathscr{C}^{1/2}, \mathscr{A}^{1/2}\mathscr{B}^{1/2}\mathscr{C}^{1/2}C\}; \quad C^2=I \tag{7}$$

$$\{2_1\}\otimes\{\mathscr{I}\}=\{I, \mathscr{C}^{1/2}C, \mathscr{A}^{1/2}\mathscr{B}^{1/2}\mathscr{C}^{1/2}, \mathscr{A}^{1/2}\mathscr{B}^{1/2}C\}; \quad C^2=I \tag{8}$$

where $\{\mathscr{T}\}$ has been omitted—so allowing us to write

$$\mathscr{A}^{1/2}\mathscr{B}^{1/2}\mathscr{C}^{1/2}.\mathscr{C}^{1/2}C=\mathscr{A}^{1/2}\mathscr{B}^{1/2}\mathscr{C}C \Rightarrow \mathscr{A}^{1/2}\mathscr{B}^{1/2}C, \text{ etc.} \tag{9}$$

Now $\mathscr{A}^{1/2}\mathscr{B}^{1/2}C=(\tfrac{1}{4}\tfrac{1}{4}0/001)^{1/2}$, which suggests the geometrically meaningful transformation

$$\mathscr{A}^{1/2}\mathscr{B}^{1/2}C=\mathfrak{C}, \quad \text{i.e.} \quad C=\mathscr{A}^{-1/2}\mathscr{B}^{-1/2}\mathfrak{C} \Rightarrow \mathscr{A}^{1/2}\mathscr{B}^{1/2}\mathfrak{C}, \tag{10}$$

for converting (8) into (7). It readily follows that

$$\{2_1\}\otimes\{\mathscr{I}\}=\{I, \mathscr{C}^{1/2}\mathscr{A}^{1/2}\mathscr{B}^{1/2}\mathfrak{C}, \mathscr{A}^{1/2}\mathscr{B}^{1/2}\mathscr{C}^{1/2}, \mathfrak{C}\}, \tag{11}$$

which is seen to be identical with (7) except that C has been replaced by $\mathfrak{C}$.

To prove that $\{3_1\}\otimes\{\mathscr{R}\}=\{3\}\otimes\{\mathscr{R}\}$, we display these as

$$\{3\}\otimes\{\mathscr{R}\}=\left\{\begin{matrix} I & C & C^2 \\ \mathscr{A}^{2/3}\mathscr{B}^{1/3}\mathscr{C}^{1/3}, & \mathscr{A}^{2/3}\mathscr{B}^{1/3}\mathscr{C}^{1/3}C, & \mathscr{A}^{2/3}\mathscr{B}^{1/3}\mathscr{C}^{1/3}C^2 \\ \mathscr{A}^{1/3}\mathscr{B}^{2/3}\mathscr{C}^{2/3}, & \mathscr{A}^{1/3}\mathscr{B}^{2/3}\mathscr{C}^{2/3}C, & \mathscr{A}^{1/3}\mathscr{B}^{2/3}\mathscr{C}^{2/3}C^2 \end{matrix}\right\} \tag{12}$$

$$\{3_1\} \otimes \{\mathscr{R}\} = \left\{ \begin{matrix} I & \mathscr{C}^{1/3}C & \mathscr{C}^{2/3}C^2 \\ \mathscr{A}^{2/3}\mathscr{B}^{1/3}\mathscr{C}^{1/3}, & \mathscr{A}^{2/3}\mathscr{B}^{1/3}\mathscr{C}^{2/3}C, & \mathscr{A}^{2/3}\mathscr{B}^{1/3}C^2 \\ \mathscr{A}^{1/3}\mathscr{B}^{2/3}\mathscr{C}^{2/3}, & \mathscr{A}^{1/3}\mathscr{B}^{2/3}C, & \mathscr{A}^{1/3}\mathscr{B}^{2/3}\mathscr{C}^{1/3}C^2 \end{matrix} \right\} \tag{13}$$

where $\mathscr{A}^{1/3}\mathscr{B}^{2/3}C = (\frac{1}{6}\frac{1}{6}0/001)^{1/3}_{+}$ refers to the doubly-centred hexagonal cell (Fig. 7.2). This suggests the transformation

$$\left. \begin{aligned} \mathscr{A}^{1/3}\mathscr{B}^{2/3}C = C, \text{ i.e. } \mathfrak{C} &= \mathscr{A}^{-1/3}\mathscr{B}^{-2/3}\mathfrak{C} \Rightarrow \mathscr{A}^{2/3}\mathscr{B}^{1/3}\mathfrak{C} \\ C^2 = (\mathscr{A}^{2/3}\mathscr{B}^{1/3}\mathfrak{C})^2 &\Rightarrow \mathscr{A}^{1/3}\mathscr{B}^{2/3}\mathfrak{C}^2 \end{aligned} \right\} \tag{14}$$

for converting (13) into (12). It readily follows that

$$\{3_1\} \otimes \{\mathscr{R}\} = \left\{ \begin{matrix} I, & \mathscr{C}^{1/3}\mathscr{A}^{2/3}\mathscr{B}^{1/3}\mathfrak{C}, & \mathscr{C}^{2/3}\mathscr{A}^{1/3}\mathscr{B}^{2/3}\mathfrak{C}^2 \\ \mathscr{A}^{2/3}\mathscr{B}^{1/3}\mathscr{C}^{1/3}, & \mathscr{A}^{1/3}\mathscr{B}^{2/3}\mathscr{C}^{2/3}\mathfrak{C}, & \mathfrak{C}^2 \\ \mathscr{A}^{1/3}\mathscr{B}^{2/3}\mathscr{C}^{2/3}, & \mathfrak{C}, & \mathscr{A}^{2/3}\mathscr{B}^{1/3}\mathscr{C}^{1/3}\mathfrak{C}^2 \end{matrix} \right\} \tag{15}$$

which is seen to be identical with (12) except that C has been replaced by $\mathfrak{C}$.

To prove that $\{2gg\} \otimes \{\mathscr{E}\} = \{2gc\} \otimes \{\mathscr{E}\}$, we display these as

$$\{2gg\} \otimes \{\mathscr{E}\} = \left\{ \begin{matrix} I, (001)^{1/2}, & \mathscr{A}^{1/2}\mathscr{B}^{1/2}\mathscr{C}^{1/2}(010)_m, & \mathscr{A}^{1/2}\mathscr{B}^{1/2}\mathscr{C}^{1/2}(100)_m \\ \mathscr{A}^{1/2}\mathscr{B}^{1/2}, & \mathscr{A}^{1/2}\mathscr{B}^{1/2}(001)^{1/2}, & \mathscr{C}^{1/2}(010)_m, & \mathscr{C}^{1/2}(100)_m \end{matrix} \right\} \tag{16}$$

$$\{2gc\} \otimes \{\mathscr{E}\} = \left\{ \begin{matrix} I, & (001)^{1/2}, & \mathscr{C}^{1/2}\mathscr{A}^{1/2}(010)_m, & \mathscr{C}^{1/2}\mathscr{A}^{1/2}(100)_m \\ \mathscr{A}^{1/2}\mathscr{B}^{1/2}, & \mathscr{A}^{1/2}\mathscr{B}^{1/2}(001)^{1/2}, & \mathscr{B}^{1/2}\mathscr{C}^{1/2}(010)_m, & \mathscr{B}^{1/2}\mathscr{C}^{1/2}(100)_m \end{matrix} \right\}, \tag{17}$$

and introduce the transformation

$$\mathscr{B}^{1/2}(010)_m = (010)^{\star}_m, \quad \mathscr{A}^{1/2}(100)_m = (100)^{\star}_m. \tag{18}$$

Substituting from (18) into (17) converts (17) into (16) with $(010)_m$, $(100)_m$ replaced by $(010)^{\star}_m$, $(100)^{\star}_m$ respectively.

A counter-example is provided by $\{a\} \otimes \{\mathscr{I}\} \neq \{m\} \otimes \{\mathscr{I}\}$. Thus, displaying these as

$$\{m\} \otimes \{\mathscr{I}\} = \{I, (001)_m, \mathscr{A}^{1/2}\mathscr{B}^{1/2}\mathscr{C}^{1/2}, \mathscr{A}^{1/2}\mathscr{B}^{1/2}\mathscr{C}^{1/2}(001)_m\} \tag{19}$$

$$\{a\} \otimes \{\mathscr{I}\} = \{I, \mathscr{A}^{1/2}(001)_m, \mathscr{A}^{1/2}\mathscr{B}^{1/2}\mathscr{C}^{1/2}, \mathscr{B}^{1/2}\mathscr{C}^{1/2}(001)_m\} \tag{20}$$

we see that the only formal possibilities for a transformation are either

$$\mathscr{A}^{1/2}(001)_m = (001)^{\star}_m \text{ or } \mathscr{B}^{1/2}\mathscr{C}^{1/2}(001)_m = (001)^{\star}_m. \tag{21}$$

However, neither of these is admissible since a glide reflection ranks as inherently distinct from a pure reflection.

The independent centred space groups constructed by applications of Theorem (2) are listed in Table 13.1.

Table 13.1—Independent Centred Space Groups

monoclinic	I	2; m, c; $\dfrac{2}{m}$, $\dfrac{2}{c}$
orthorhombic	C	$2mm$, $2mc$, $2cc$; 222, 222_1; $\dfrac{2}{m}\dfrac{2}{m}\dfrac{2}{m}$, $\dfrac{2}{m}\dfrac{2}{m}\dfrac{2}{a}$, $\dfrac{2}{m}\dfrac{2}{c}\dfrac{2}{m}$, $\dfrac{2}{c}\dfrac{2}{c}\dfrac{2}{m}$, $\dfrac{2}{c}\dfrac{2}{c}\dfrac{2}{a}$
	A	$2mm$, $2na$, $2bm$, $2ba$
	I	$2mm$, $2na$, $2ba$; 222, $2_12_12_1$; $\dfrac{2}{m}\dfrac{2}{m}\dfrac{2}{m}$, $\dfrac{2}{m}\dfrac{2}{m}\dfrac{2}{a}$, $\dfrac{2}{m}\dfrac{2}{a}\dfrac{2}{a}$, $\dfrac{2}{b}\dfrac{2}{c}\dfrac{2}{a}$
	F	$2mm$; 222; $\dfrac{2}{m}\dfrac{2}{m}\dfrac{2}{m}$
tetragonal	I	4, 4_1; $\bar{4}$; $\dfrac{4}{m}$; 422, 4_122; $4mm$, $4cm$; $\bar{4}2m$, $\bar{4}n2$, $\bar{4}c2$; $\dfrac{4}{m}\dfrac{2}{m}\dfrac{2}{m}$, $\dfrac{4}{m}\dfrac{2}{c}\dfrac{2}{m}$
hexagonal	R	3; $\bar{3}$; 32; $3m$, $3c$; $\bar{3}\dfrac{2}{m}$, $\bar{3}\dfrac{2}{c}$
cubic	I	23, 2_13; 432, 4_132; $\dfrac{2}{m}\bar{3}$, $\dfrac{2}{a}\bar{3}$; $\dfrac{4}{m}\bar{3}\dfrac{2}{m}$, $\dfrac{4}{a}\bar{3}\dfrac{2}{m}$; $\bar{4}3m$
	F	23; 432, 4_132; $\dfrac{2}{m}\bar{3}$, $\dfrac{2}{a}\bar{3}$; $\dfrac{4}{m}\bar{3}\dfrac{2}{m}$, $\dfrac{4}{m}\bar{3}\dfrac{2}{c}$; $\bar{4}3m$, $\bar{4}3c$

This table comprises 68 independent space groups, of which 36 are symmorphic and the remainder non-symmorphic.

13.3 THE SPACE GROUP $\left\{\dfrac{4_1}{a}\right\} \otimes \{\mathscr{I}\}$

The set

$$\left\{\dfrac{4_1}{a}\right\} = \{4_1\} + \mathscr{A}^{1/2}J\{4_1\} = \{4_1\} + \mathscr{B}^{1/2}M\{4_1\} \text{ mod } \{\mathscr{I}\}$$

$$= \left\{ \begin{array}{llll} I, & \mathscr{C}^{1/4}C, & \mathscr{C}^{2/4}C^2, & \mathscr{C}^{3/4}C^3 \\ \mathscr{A}^{1/2}J, & \mathscr{A}^{1/2}\mathscr{C}^{-1/4}JC, & \mathscr{A}^{1/2}\mathscr{C}^{-2/4}JC^2, & \mathscr{A}^{1/2}\mathscr{C}^{-3/4}JC^3 \end{array} \right\}, \qquad (23)$$

where

$$C = (001)^{1/4}; \quad C\mathscr{A}^{\lambda}\mathscr{B}^{\mu}\mathscr{C}^{\nu}C^{-1} = \mathscr{A}^{-\mu}\mathscr{B}^{\lambda}\mathscr{C}^{\nu}, \qquad (24)$$

is not a groupoid with respect to $\{\mathscr{T}\}$, as can be seen either directly on symmetry grounds or from the product

$$\mathscr{C}^{1/4}C.\mathscr{A}^{1/2}J = \mathscr{C}^{1/4}.C\mathscr{A}^{1/2}.J = \mathscr{C}^{1/4}.\mathscr{B}^{1/2}C.J$$

$$= \mathscr{B}^{1/2}\mathscr{C}^{1/4}JC \neq \mathscr{A}^{1/2}\mathscr{C}^{-1/4}JC \text{ mod } \{\mathscr{T}\}. \qquad (25)$$

However, this product belongs to the enlarged set

$$\left\{\dfrac{4_1}{a}\right\} \otimes \{\mathscr{T}, \mathscr{A}^{1/2}\mathscr{B}^{1/2}\mathscr{C}^{1/2}\mathscr{T}\} \text{ i.e. } \left\{\dfrac{4_1}{a}\right\} \otimes \{\mathscr{I}\}, \qquad (26)$$

by virtue of

$$\mathscr{A}^{1/2}\mathscr{B}^{1/2}\mathscr{C}^{1/2}.\mathscr{A}^{1/2}\mathscr{C}^{-1/4}JC \Rightarrow \mathscr{B}^{1/2}\mathscr{C}^{1/4}JC. \qquad (27)$$

More generally, utilising an obvious notation, we find

$$\mathscr{A}^{1/2}J\{4_1\}.\{4_1\} = \mathscr{A}^{1/2}J\{4_1\} \text{ mod } \{\mathscr{T}\},$$

$$\{4_1\}.\mathscr{A}^{1/2}J\{4_1\} = \mathscr{A}^{1/2}J\{4_1\}\{4_1\} \text{ mod } \{\mathscr{I}\} = \mathscr{A}^{1/2}J\{4_1\} \text{ mod } \{\mathscr{I}\}$$

which results may be usefully summarised by writing

$$\left\{\dfrac{4_1}{a}\right\} . \left\{\dfrac{4_1}{a}\right\} = \left\{\dfrac{4_1}{a}\right\} \text{ mod } \{\mathscr{I}\}. \qquad (28)$$

Coupling this with the further results

$$\left\{\dfrac{4_1}{a}\right\} . \mathscr{A}^{1/2}\mathscr{B}^{1/2}\mathscr{C}^{1/2} = \mathscr{A}^{1/2}\mathscr{B}^{1/2}\mathscr{C}^{1/2} \left\{\dfrac{4_1}{a}\right\} \text{ mod } \{\mathscr{T}\}, \text{ etc.}$$

we see that $\left\{\dfrac{4_1}{a}\right\}\otimes\{\mathscr{I}\}$ constitutes a space group even though $\left\{\dfrac{4_1}{a}\right\}\otimes\{\mathscr{T}\}$ does not qualify.

13.4 DIAMOND GLIDE

Diamond glide operators have the forms

$$\mathscr{A}^{1/4}\mathscr{B}^{1/4}(001)_m\,,\;\mathscr{A}^{1/4}\mathscr{B}^{1/4}\mathscr{C}^{1/4}(1\bar{1}0)_m\,,\;\text{etc.,} \tag{29}$$

which are characterised by powers

$$[\mathscr{A}^{1/4}\mathscr{B}^{1/4}(001)_m]^2=\mathscr{A}^{1/2}\mathscr{B}^{1/2},\;[\mathscr{A}^{1/4}\mathscr{B}^{1/4}\mathscr{C}^{1/4}(1\bar{1}0)_m]^2=\mathscr{A}^{1/2}\mathscr{B}^{1/2}\mathscr{C}^{1/2},\;\text{etc.} \tag{30}$$

Clearly such operators could only be associated with centred translation groups. Thus, introducing the groupoid

$$\{2dd\}=\{2\}+\mathscr{A}^{1/4}\mathscr{B}^{1/4}\mathscr{C}^{1/4}(010)_m\{2\} \tag{31}$$
$$=\{I,\,(001)^{1/2},\,\mathscr{A}^{1/4}\mathscr{B}^{1/4}\mathscr{C}^{1/4}(010)_m,\,\mathscr{A}^{1/4}\mathscr{B}^{1/4}\mathscr{C}^{1/4}(100)_m\},$$

where

$$\mathscr{A}^{1/4}\mathscr{B}^{1/4}\mathscr{C}^{1/4}(010)_m=\mathscr{A}^{1/4}\mathscr{C}^{1/4}(0\tfrac{1}{8}0/010)_m\,,\;\mathscr{A}^{1/4}\mathscr{B}^{1/4}\mathscr{C}^{1/4}(100)_m$$
$$=\mathscr{B}^{1/4}\mathscr{C}^{1/4}(\tfrac{1}{8}00/100)_m, \tag{32}$$

we find

$$\{2dd\}.\{2dd\}=\{2dd\}\;\text{mod}\;\{\mathscr{T},\,\mathscr{A}^{1/2}\mathscr{B}^{1/2}\mathscr{T},\,\mathscr{B}^{1/2}\mathscr{C}^{1/2}\mathscr{T},\,\mathscr{C}^{1/2}\mathscr{A}^{1/2}\mathscr{T}\},$$
$$\text{i.e. }\{2dd\}\;\text{mod}\;\{\mathscr{F}\},$$
$$\{2dd\}.\mathscr{A}^{1/2}\mathscr{B}^{1/2}=\mathscr{A}^{1/2}\mathscr{B}^{1/2}\{2dd\}\;\text{mod}\;\{\mathscr{F}\},\;\text{etc.,}$$

showing that $\{2dd\}\otimes\{\mathscr{F}\}$ is a space group. This is one of the two face-centred orthorhombic diamond space groups. The companion space group is sufficiently defined by

$$\left\{\dfrac{2\,2\,2}{d\,d\,d}\right\}=\{222\}+\mathscr{A}^{1/4}\mathscr{B}^{1/4}\mathscr{C}^{1/4}J\{222\}, \tag{33}$$

being readily understandable by comparison with $\{2gg\}$—see (30), Chapter 12.

There exist five body-centred tetragonal diamond space groups, sufficiently defined by the groupoids:

$$\{4_1md\}=\{4_1\}+\mathscr{B}^{1/2}(010)_m\{4_1\} \tag{34}$$
$$\{4_1cd\}=\{4_1\}+\mathscr{B}^{1/2}\mathscr{C}^{1/2}(010)_m\{4_1\} \tag{35}$$

$$\{\bar{4}2d\} = \{\bar{4}\} + \mathscr{B}^{1/2}\mathscr{C}^{1/4}(100)^{1/2}\{\bar{4}\} \tag{36}$$

$$\left\{\frac{4_1}{a}\frac{2}{m}\frac{2}{d}\right\} = \{4_122\} + \mathscr{A}^{1/2}J\{4_122\} \tag{37}$$

$$\left\{\frac{4_1}{a}\frac{2}{c}\frac{2}{d}\right\} = \{4_122\} + \mathscr{A}^{1/2}\mathscr{C}^{1/2}J\{4_122\}. \tag{38}$$

By reference to $\{4mm\}$, (38), Chapter 12, we see that $\{4_1md\}$ includes the operators

$$\mathscr{B}^{1/2}(010)_m \,, \ \mathscr{B}^{1/2}\mathscr{C}^{2/4}(100)_m \Rightarrow \mathscr{A}^{1/2}(100)_m \tag{39}$$

which signify pure reflections, and the operators

$$\left.\begin{aligned}
&\mathscr{B}^{1/2}\mathscr{C}^{1/4}(\bar{1}10)_m = \mathscr{A}^{1/4}\mathscr{B}^{1/4}\mathscr{C}^{1/4}\cdot\mathscr{A}^{-1/4}\mathscr{B}^{1/4}(\bar{1}10)_m = \\
&\qquad \mathscr{A}^{1/4}\mathscr{B}^{1/4}\mathscr{C}^{1/4}(\tfrac{1}{8}\tfrac{1}{8}0/\bar{1}10)_m \\
&\mathscr{B}^{1/2}\mathscr{C}^{-1/4}(110)_m = \mathscr{A}^{-1/4}\mathscr{B}^{1/4}\mathscr{C}^{-1/4}\cdot\mathscr{A}^{1/4}\mathscr{B}^{1/4}(110)_m = \\
&\qquad \mathscr{A}^{-1/4}\mathscr{B}^{1/4}\mathscr{C}^{-1/4}(\tfrac{1}{8}\tfrac{1}{8}0/110)_m
\end{aligned}\right\} \tag{40}$$

which are seen to be diamond glide reflections. The corresponding operators in $\{4_1cd\}$ are

$$\mathscr{B}^{1/2}\mathscr{C}^{1/2}(010)_m \,, \ \mathscr{B}^{1/2}(100)_m \Rightarrow \mathscr{A}^{1/2}\mathscr{C}^{1/2}(100)_m \tag{39a}$$

$$\mathscr{B}^{1/2}\mathscr{C}^{3/4}(\bar{1}10)_m \,, \ \mathscr{B}^{1/2}\mathscr{C}^{1/4}(110)_m \tag{40a}$$

which signify axial and diamond glides respectively. As regards $\{\bar{4}2d\}$, reference to $\{\bar{4}2m\}$, (49), Chapter 12, indicates a pair of 2-fold rotation operators

$$\mathscr{B}^{1/2}\mathscr{C}^{1/4}(100)^{1/2} = (0\tfrac{1}{4}\tfrac{1}{8}/100)^{1/2}, \ \mathscr{B}^{1/2}\mathscr{C}^{1/4}(010)^{1/2} \Rightarrow \mathscr{A}^{1/2}\mathscr{C}^{3/4}(010)^{1/2}$$
$$= (\tfrac{1}{4}0\tfrac{3}{8}/010)^{1/2}, \tag{41}$$

accompanied by a pair of diamond glide operators

$$\mathscr{B}^{1/2}\mathscr{C}^{1/4}(\bar{1}10)_m \text{—see (40) above, } \mathscr{B}^{1/2}\mathscr{C}^{1/4}(110)_m \text{—see (40a) above.} \tag{42}$$

Most of the operators of $\left\{\dfrac{4_1}{a}\dfrac{2}{m}\dfrac{2}{d}\right\}$ have already been covered in $\left\{\dfrac{4_1}{a}\right\}$ and $\{4_122\}$, and there remain only the following four for consideration:

$$\mathscr{A}^{1/2}(100)_m \,, \ \mathscr{A}^{1/2}\mathscr{C}^{2/4}(010)_m \Rightarrow \mathscr{B}^{1/2}(010)_m \tag{43}$$

which signify pure reflections, and

$$\mathscr{A}^{1/2}\mathscr{C}^{1/4}(\bar{1}10)_m \Rightarrow \mathscr{B}^{1/2}\mathscr{C}^{3/4}(\bar{1}10)_m, \; \mathscr{A}^{1/2}\mathscr{C}^{3/4}(110)_m \Rightarrow \mathscr{B}^{1/2}\mathscr{C}^{1/4}(110)_m, \qquad (44)$$

which—see (40a)—signify diamond glides. Replacing $\mathscr{A}^{1/2}$ by $\mathscr{A}^{1/2}\mathscr{C}^{1/2}$ converts the reflection operators (43) into axial glides, whilst still preserving the diamond character of the operators (44)—see (40), so accounting for the groupoid $\left\{\dfrac{4_1\,2\,2}{a\;c\,d}\right\}$.

That $\{4_1md\}\otimes\{\mathscr{I}\}$, etc., form space groups may be proved by the systematic computation of subgroupoid products in the usual way. All the diamond space groups are listed in Table 13.2.

Table 13.2—Diamond and Allied Groupoids

Crystal system	Lattice type	Groupoid
orthorhombic	F	$2dd, \dfrac{2\,2\,2}{d\,d\,d}$
tetragonal	I	$4_1/a, \bar{4}2d$
		$4_1md, 4_1cd, \dfrac{4_1}{a}\dfrac{2}{m}\dfrac{2}{d}, \; \dfrac{4_1}{a}\dfrac{2\,2}{c\,d}$
cubic	I	$\bar{4}3d, \; \dfrac{4_1}{a}\bar{3}\dfrac{2}{d}$
	F	$\dfrac{2}{d}\bar{3}, \; \dfrac{4_1}{d}\bar{3}\dfrac{2}{m}, \; \dfrac{4_1}{d}\bar{3}\dfrac{2}{c}$

13.5 CUBIC SYSTEMS

There exist two body-centred cubic and three face-centred cubic diamond space groups, defined in increasing order of complexity as follows:

$$\mathscr{F}: \left\{\dfrac{2}{d}\bar{3}\right\} = \{23\} + \mathscr{A}^{1/4}\mathscr{B}^{1/4}\mathscr{C}^{1/4}J\{23\} \qquad (45)$$

$$\mathscr{I}: \{\bar{4}3d\} = \{2_13\} + \mathscr{A}^{1/4}\mathscr{B}^{1/4}\mathscr{C}^{-1/4}J(100)^{1/4}\{2_13\} \qquad (46)$$

$$\mathscr{I}: \left\{\dfrac{4_1}{a}\bar{3}\dfrac{2}{d}\right\} = \{4_132\} + J\{4_132\} \qquad (47)$$

$$\mathscr{F}: \left\{\dfrac{4_1}{d}\bar{3}\dfrac{2}{m}\right\} = \{4_132\} + \mathscr{A}^{-1/4}\mathscr{B}^{-1/4}\mathscr{C}^{-1/4}J\{4_132\} \qquad (48)$$

$$\mathscr{F}: \left\{\frac{4_1}{d}\,\bar{3}\,\frac{2}{c}\right\} = \{4_132\} + \mathscr{A}^{1/4}\mathscr{B}^{1/4}\mathscr{C}^{1/4}J\{4_132\}. \tag{49}$$

The groupoid $\left\{\frac{2}{d}\bar{3}\right\}$ may be readily understood by comparison with $\left\{\frac{2}{g}\bar{3}\right\}$, (87), Chapter 12. This contains the subgroupoid $\left\{\frac{2}{g}\right\}$ which clearly now becomes $\left\{\frac{2}{d}\right\}$. It also contains $\{\{\mathscr{A}^{1/2}\mathscr{B}^{1/2}\mathscr{C}^{1/2}J(111)^{1/3}\}\}_6$, which now becomes $\{\{\mathscr{A}^{1/4}\mathscr{B}^{1/4}\mathscr{C}^{1/4}J(111)^{1/3}\}\}_6$ providing an alternative realisation of $\{\bar{3}\}$. Finally we note the powers

$$[\mathscr{A}^{1/4}\mathscr{B}^{1/4}\mathscr{C}^{1/4}(100)_m]^2 = \mathscr{B}^{1/2}\mathscr{C}^{1/2}, \text{ etc.,} \tag{50}$$

indicating $\{\mathscr{F}\}$ as the appropriate centred translation group. Similarly, $\{\bar{4}3d\}$ may be understood by comparison with $\{4_132\}$, (30), Chapter 12. They both contain $\{\{(111)^{1/3}\}\}_3$, etc. Also, $\{4_132\}$ contains $\{\{\mathscr{A}^{1/4}\mathscr{B}^{1/4}\mathscr{C}^{-1/4}(100)^{1/4}\}\}_4$, which now becomes $\{\{\mathscr{A}^{1/4}\mathscr{B}^{1/4}\mathscr{C}^{-1/4}J(100)^{1/4}\}\}_4$ so providing a realisation of $\{\bar{4}\}$. Finally, the 2-fold rotation operator $\mathscr{A}^{1/4}\mathscr{B}^{1/4}\mathscr{C}^{-1/4}(101)^{1/2}$ of $\{4_132\}$ becomes the diamond glide operator $\mathscr{A}^{1/4}\mathscr{B}^{1/4}\mathscr{C}^{-1/4}(101)_m$, etc., of $\{\bar{4}3d\}$, as may be seen from the fact that $[11\bar{1}] \,\|\, (101)$. It follows that

$$[\mathscr{A}^{1/4}\mathscr{B}^{1/4}\mathscr{C}^{-1/4}(101)_m]^2 = \mathscr{A}^{1/2}\mathscr{B}^{1/2}\mathscr{C}^{-1/2}, \text{ etc.,} \tag{51}$$

so indicating $\{\mathscr{I}\}$ as the appropriate centred space group.

Dealing now with $\left\{\frac{4_1}{a}\bar{3}\,\frac{2}{d}\right\}$, this is seen to include $\left\{\frac{2_1}{a}\bar{3}\right\}$, (90) Chapter 12. Also it includes

$$\left\{\frac{4_1}{a}\right\} = \{\{\mathscr{A}^{1/4}\mathscr{B}^{1/4}\mathscr{C}^{-1/4}(100)^{1/4}\}\}_4 + J\{\{''\}\}_4, \tag{52}$$

where $\{\{''\}\}_4$ originates from $\{4_132\}$ as noted above, and

$$J\{\{''\}\}_4 = \mathscr{A}^{1/2}\mathscr{B}^{1/2}(100)_m\{\{''\}\}_4 \tag{53}$$

as follows from a detailed analysis, e.g.

$$J . \mathscr{A}^{1/4}\mathscr{B}^{1/4}\mathscr{C}^{-1/4}(100)^{1/4} = \mathscr{A}^{1/2}\mathscr{B}^{-1/2}(100)_m . \mathscr{A}^{3/4}\mathscr{B}^{1/4}\mathscr{C}^{1/4}(100)^{3/4}$$

$$\Rightarrow \mathscr{A}^{1/2}\mathscr{B}^{1/2}(100)_m[\mathscr{A}^{1/4}\mathscr{B}^{1/4}\mathscr{C}^{-1/4}(100)^{1/4}]^3, \tag{54}$$

on bearing in mind $J = (100)_m(100)^{1/2}$. It also includes the operators

$$\mathscr{A}^{1/4}\mathscr{B}^{1/4}\mathscr{C}^{-1/4}(101)^{1/2}, \; J\mathscr{A}^{1/4}\mathscr{B}^{1/4}\mathscr{C}^{-1/4}(101)^{1/2} \text{ i.e. } \mathscr{A}^{-1/4}\mathscr{B}^{-1/4}\mathscr{C}^{1/4}(101)_m \tag{55}$$

which originate respectively from $\{4_1 32\}$, $J\{4_1 32\}$, so providing a 2-fold rotation axis accompanied by a diamond glide reflection. This satisfies (50), indicating $\{\mathscr{I}\}$ as the appropriate centred space group.

The groupoid $\left\{\dfrac{4_1}{d}\bar{3}\dfrac{2}{m}\right\}$ is comparable with $\left\{\dfrac{4_1}{a}\bar{3}\dfrac{2}{d}\right\}$, but associated with $\{\mathscr{F}\}$ instead of with $\{\mathscr{I}\}$. In place of (53) it contains

$$\mathscr{A}^{-1/4}\mathscr{B}^{-1/4}\mathscr{C}^{-1/4}.\mathscr{A}^{1/2}\mathscr{B}^{1/2}(100)_m\{\{''\}\}_4 = \mathscr{A}^{1/4}\mathscr{B}^{1/4}\mathscr{C}^{-1/4}(100)_m\{\{\ \}\}_4, \quad (56)$$

which signifies a diamond glide on (100) so that $\left\{\dfrac{4_1}{a}\right\}$ above now becomes $\left\{\dfrac{4_1}{d}\right\}$. Also, the second operator of (55) becomes

$$\mathscr{A}^{-1/4}\mathscr{B}^{-1/4}\mathscr{C}^{-1/4}.\mathscr{A}^{-1/4}\mathscr{B}^{-1/4}\mathscr{C}^{1/4}(101)_m = \mathscr{A}^{-1/2}\mathscr{B}^{-1/2}(101)_m \Rightarrow (101)_m, \quad (57)$$

i.e. a pure reflection replacing the diamond glide so that $\left\{\dfrac{2}{d}\right\}$ above now becomes $\left\{\dfrac{2}{m}\right\}$. Finally, it includes

$$\{\{\mathscr{A}^{-1/4}\mathscr{B}^{-1/4}\mathscr{C}^{-1/4}J(111)^{1/3}\}\}_6, \text{ etc.,} \quad (58)$$

providing yet another realisation of $\{\bar{3}\}$. Since

$$[\mathscr{A}^{1/4}\mathscr{B}^{1/4}\mathscr{C}^{-1/4}(100)_m]^2 = \mathscr{B}^{1/2}\mathscr{C}^{-1/2}, \text{ etc.,} \quad (59)$$

we arrive at the space group $\left\{\dfrac{4_1}{d}\bar{3}\dfrac{2}{m}\right\}\otimes\{\mathscr{F}\}$ characteristic of the diamond crystal structure. This appears macroscopically as the point-group symmetry $\left\{\dfrac{4}{m}\bar{3}\dfrac{2}{m}\right\}$, owing to the suppression of $\{\mathscr{F}\}$ and of the translation components in glide reflection.

The final groupoid $\left\{\dfrac{4_1}{d}\bar{3}\dfrac{2}{c}\right\}$ differs only slightly from $\left\{\dfrac{4_1}{d}\bar{3}\dfrac{2}{m}\right\}$. In place of (56) we find

$$\mathscr{A}^{1/2}\mathscr{B}^{1/2}\mathscr{C}^{1/2}.\mathscr{A}^{1/4}\mathscr{B}^{1/4}\mathscr{C}^{-1/4}(100)_m\{\{''\}\}_4 \to \mathscr{A}^{-1/4}\mathscr{B}^{-1/4}\mathscr{C}^{1/4}(100)_m\{\{''\}\}_4,$$

which again signifies a diamond glide on (100). Also, (58) becomes

$$\{\{\mathscr{A}^{1/4}\mathscr{B}^{1/4}\mathscr{C}^{1/4}J(111)^{1/3}\}\}_6$$

i.e. again a realisation of $\{\bar{3}\}$ as in $\left\{\dfrac{2}{d}\bar{3}\right\}$. However, (57) becomes

$$\mathscr{A}^{1/2}\mathscr{B}^{1/2}\mathscr{C}^{1/2}(101)_m = \mathscr{B}^{1/2}(\tfrac{1}{4}0\tfrac{1}{4}/101)_m \tag{60}$$

which signifies an axial glide on (101) replacing the previous pure reflection.

13.6 SUMMARY

This completes our analysis of the classical space groups. There are 230 of these, classified as in Table 13.3. Columns 1 and 2 refer to the 73(37 + 36) space groups, usually termed symmorphic space groups, which do not involve screw axes or glide planes. However, we have also termed them Bravais space groups since they would have been understood and accepted by Bravais. Column 3 refers to the 112 non-symmorphic space groups which exist in primitive Bravais space lattices. All these could also exist in appropriate centred Bravais space lattices, but only 32 prove to be independent as indicated in column 4. Column 5 refers to the 13 non-symmorphic space groups which have no counterparts in primitive space lattices, comprising 12 of the diamond type plus the exceptional tetragonal space group $I4_1/a$.

There are 17 two-dimensional space groups. These may be deduced directly (e.g. see [9]). However they may also be obtained as the three-dimensional space groups which generate plane patterns. This theory is outlined in Appendix 7.

Table 13.3—Breakdown of the 230 Classical Space Groups

	PB	CB	PN	CN	D	
triclinic	2	—	—	—	—	2
monoclinic	3	3	5	2	—	13
orthorhombic	3	10	27	17	2	59
tetragonal	8	8	41	5	6	68
cubic	5	10	10	6	5	36
hexagonal	16	5*	29	2*	—	52
	37	36	112	32	13	= 230

PB: primitive Bravais (i.e. symmorphic)   PN: primitive non-Bravais
CB: centred Bravais   CN: centred non-Bravais
  D: diamond glide $+I4_1/a$
  * alternatively classified as primitive rhombohedral

Chapter 14

The Colour Space Groups

14.1 COLOUR SPACE GROUPS: BRAVAIS TYPE

As seen in Chapter 9, symmorphic (Bravais) space groups are of the form $\{G\} \otimes \{\mathcal{T}\}$, $\{G\} \otimes \{\mathcal{I}\}$, etc., subject to the coupling conditions (11)–(15), Chapter 8, as appropriate. This suggests that we may construct Bravais colour space groups of the following forms:

(I) $\{G'\} \otimes \{\mathcal{T}\}$, $\{G'\} \otimes \{\mathcal{I}\}$, etc. where $\{G'\} \Leftrightarrow \{G\}$, i.e. the direct product of a colour point group (see Chapter 4) with an ordinary translation group. Thus, for instance, $\{4'32'\} \Leftrightarrow \{432\}$ so providing the colour space groups $\{4'32'\} \otimes \{\mathcal{I}\}$, $\{4'32'\} \otimes \{\mathcal{T}\}$. Proceeding systematically through Table 4.1, and bearing in mind the left-hand columns of Table 9.1, we arrive at the 116 type I colour space groups listed in Table 14.1.

(II) $\{G\} \otimes \{I, S\,\mathscr{C}^{1/2}\}\,\{\mathcal{T}\}$, etc., i.e. the direct product of an ordinary point group with a colour translation group. Accordingly, combining Table 8.1 systematically with Table 1.2, we arrive at the 96 type II colour space groups listed in Table 14.2.

Certain ambiguities of setting arise with the colour space groups as for the ordinary space groups. For instance, the symmetries $4'mm'$, $4'm'm$ are equivalent and designated $4'mm'$ in Table 14.1. However, relative to the (001) face of the primitive tetragonal cell, they yield two independent settings (Fig. 14.1). This means that the space groups P $4'mm'$, I $4'mm'$ in Table 14.1 must be supplemented by the space groups P $4'm'm$, I $4'm'm$. Similar considerations for other symmetries lead to the thirty-two supplementary type I colour space groups listed in Table 14.1a. However, a counter-example occurs with R $\bar{3}'\frac{2}{m'}$ (see Table 4.3), which—by contrast with P $\bar{3}'\frac{2}{m'}$, etc.—is the only colour space group in the rhombohedral system corresponding with the three colour point groups $\bar{3}'\frac{2'}{m}$, $\bar{3}\frac{2}{m'}$, $\bar{3}'\frac{2'}{m}$.

Colour space groups of the form $\{G'\} \otimes \{I, S\,\mathscr{C}^{1/2}\}\,\{\mathcal{T}\}$, etc., may be constructed i.e. the direct product of a colour point group with a colour translation group. However, these prove to be equivalent to the type I colour

Table 14.1—Bravais Type I Colour Space Groups

Crystal system	Lattice Type	Point symmetry	Number of space groups
triclinic	P	$\bar{1}'$	1
monoclinic	P, I	$2'$; m'; $\dfrac{2'}{m}, \dfrac{2}{m'}, \dfrac{2'}{m'}$.	$2 \times 5 = 10$
orthorhombic	P, I, C, F	$22'2'$; $2'mm'$, $2m'm'$; $\dfrac{2}{m'}\dfrac{2}{m'}\dfrac{2}{m'}$, $\dfrac{2}{m}\dfrac{2'}{m}\dfrac{2'}{m}$, $\dfrac{2}{m}\dfrac{2'}{m'}\dfrac{2'}{m'}$.	$4 \times 6 = 24$
tetragonal	P, I	$4'$; $\bar{4}'$; $4'22'$, $42'2'$; $4'mm'$, $4m'm'$; $\bar{4}'m2'$, $\bar{4}'m'2$, $\bar{4}'m2'$; $\dfrac{4'}{m}\dfrac{4}{m'}\dfrac{4'}{m'}$; $\dfrac{4}{m'}\dfrac{2}{m'}\dfrac{2}{m'}$, $\dfrac{4}{m}\dfrac{2'}{m}\dfrac{2'}{m}$, $\dfrac{4'}{m'}\dfrac{2}{m}\dfrac{2'}{m'}$, $\dfrac{4'}{m'}\dfrac{2'}{m}\dfrac{2}{m'}$, $\dfrac{4}{m}\dfrac{2'}{m'}\dfrac{2'}{m'}$	$2 \times 17 = 34$
hexagonal	P	$\bar{3}'$; $32'$; $3m'$; $\bar{3}\dfrac{2'}{m'}$; $\bar{3}'\dfrac{2}{m'}$; $\bar{3}'\dfrac{2'}{m}$; $\dfrac{3'}{m'}\dfrac{3}{m'}\dfrac{3'}{m}$; $\dfrac{3}{m'}m'2$, $\dfrac{3}{m'}m2'$, $\dfrac{3}{m}m'2'$; $6'$; $6'22'$, $62'2'$; $6'mm'$, $6m'm'$, $\dfrac{6'}{m'}\dfrac{6}{m'}\dfrac{6'}{m'}$ $\dfrac{6}{m'}\dfrac{2}{m'}\dfrac{2}{m'}$, $\dfrac{6}{m'}\dfrac{2'}{m}\dfrac{2'}{m}$, $\dfrac{6'}{m}\dfrac{2}{m'}\dfrac{2'}{m}$, $\dfrac{6'}{m'}\dfrac{2'}{m'}\dfrac{2}{m}$, $\dfrac{6}{m}\dfrac{2'}{m'}\dfrac{2'}{m}$	25
	R	$\bar{3}'$; $32'$; $3m'$; $\bar{3}'\dfrac{2}{m'}$	4
cubic	P, I, F	$4'32'$; $\bar{4}'3m'$; $\dfrac{2}{m'}\bar{3}'$ $\dfrac{4'}{m}\bar{3}'\dfrac{2'}{m'}$, $\dfrac{4}{m'}\bar{3}\dfrac{2}{m'}$, $\dfrac{4'}{m'}\bar{3}'\dfrac{2'}{m}$	$3 \times 6 = 18$

Total $= 116$

Table 14.1a—Supplementary Settings to Table 14.1

Crystal system	Lattice Type	Point symmetry	Number of space groups
orthorhombic	C	$2'2'2;\ \dfrac{2'}{m'}\dfrac{2'}{m'}\dfrac{2}{m},\ \dfrac{2'}{m}\dfrac{2}{m'}\dfrac{2'}{m}$	3
	A	$2'm'm,\ 2m'm',\ 2'mm'$	3
tetragonal	P, I	$4'2'2;\ 4'm'm;\ \bar{4}2'm';$ $\bar{4}'2m',\ \bar{4}'2'm;\ \dfrac{4'}{m'}\dfrac{2}{m'}\dfrac{2'}{m'},$ $\dfrac{4'}{m}\dfrac{2'}{m'}\dfrac{2}{m'}$	$2\times7=14$
hexagonal	P	$312';\ 31m';\ \bar{3}1\dfrac{2'}{m'},\ \bar{3}'1\dfrac{2}{m'};$ $\bar{3}'1\dfrac{2'}{m};\ \dfrac{3}{m'}2m',\ \dfrac{3}{m'}2'm,$ $\dfrac{3}{m}2'm';$ $6'2'2,\ 6'm'm;\ \dfrac{6'}{m'}\dfrac{2}{m}\dfrac{2'}{m'},$ $\dfrac{6'}{m}\dfrac{2'}{m}\dfrac{2}{m'}.$	12

Total $=32$

space groups already considered. Thus, letting $\mathcal{T}^{1/2}$ stand for $\mathscr{C}^{1/2}$, $\mathscr{A}^{1/2}\mathscr{B}^{1/2}$, etc., as appear in (7)–(10), Chapter 8, we have

$$\{G'\}\otimes\{I, S\,\mathcal{T}^{1/2}\}\{\mathcal{T}\} = [\{H\}+Sk\{H\}]\otimes\{I, S\,\mathcal{T}^{1/2}\}\{\mathcal{T}\}$$
$$=\{H\}\otimes\{I, k\,\mathcal{T}^{1/2}\}\{\mathcal{T}\}+S\{H\}\otimes\{k, \mathcal{T}^{1/2}\}\{\mathcal{T}\}. \tag{1}$$

Now

$$\{H\}\otimes\{k, \mathcal{T}^{1/2}\} = k\{H\}\otimes\{I, k\,\mathcal{T}^{1/2}\}$$

on bearing in mind $k^2 \subset \{H\}$, so that (1) becomes

$$[\{H\}+Sk\{H\}]\otimes\{I, k\,\mathcal{T}^{1/2}\}\{\mathcal{T}\} \text{ i.e. } \{G'\}\otimes\{I, k\,\mathcal{T}^{1/2}\}\{\mathcal{T}\}.$$

Table 14.2—Bravais Type II Colour Space Groups

Crystal system	Lattice Type	Point symmetry	Number of space groups
triclinic	P_s	1, $\bar{1}$	2
monoclinic	P_c, P_a, P_I, I_b, I_c	2, m, $\dfrac{2}{m}$	$5 \times 3 = 15$
orthorhombic	P_c, P_I, P_C, I_c, C_A, C_c, C_a, F_c	222, $2mm$, $\dfrac{2}{m}\dfrac{2}{m}\dfrac{2}{m}$	$8 \times 3 = 24$
tetragonal	P_c, P_I, P_C, I_c	4, $\bar{4}$, 422, $4mm$, $\bar{4}2m$, $\dfrac{4}{m}$, $\dfrac{4}{m}\dfrac{2}{m}\dfrac{2}{m}$	$4 \times 7 = 28$
hexagonal	P_c, R_I	3, 32, $3m$, $\bar{3}$, $\bar{3}\dfrac{2}{m}$	$2 \times 5 = 10$
	P_c	$\dfrac{3}{m}$, $\dfrac{3}{m}m2$, 6, 622, $6mm$, $\dfrac{6}{m}$, $\dfrac{6}{m}\dfrac{2}{m}\dfrac{2}{m}$	$= 7$
cubic	P_I, F_c	23, $\dfrac{2}{m}\bar{3}$, 432, $\bar{4}3m$, $\dfrac{4}{m}\bar{3}\dfrac{2}{m}$	$2 \times 5 = 10$

Total $= 96$

This differs from $\{G'\} \otimes \{\mathscr{T}\}$ only by the presence of the factor $\{I, k\,\mathscr{T}^{1/2}\}$, which generates a diatomic motif unit from a monatomic unit. However, the essential symmetry remains unaffected. For example, Fig. 14.2 depicts a two-dimensional space group generated by $\{4'\}$ on the colour-centred square lattice, starting with a monatomic motif unit. The resulting array, however,

Table 14.2a Supplementary Settings to Table 14.2

Crystal system	Lattice Type	Point symmetry		Number of space groups
orthorhombic	P_a, P_A, I_a	*2mm*		$3 \times 1 = 3$
tetragonal	P_c, P_I, P_C, P_c	$\bar{4}m2$		$4 \times 1 = 4$
hexagonal	P_c	*312, 31m,* $\bar{3}1\dfrac{2}{m}$		$1 \times 3 = 3$
		$\dfrac{3}{m}2m$		1

Total $=$ 11

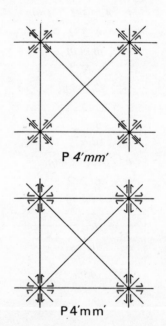

Fig. 14.1—Patterns conforming to the space groups P4′m′m. P4′mm′,

may also be generated by {4′} on the primitive square lattice, starting with an appropriate diatomic motif unit (Fig. 14.2). Accordingly the symmetry $P_c4′$ is essentially equivalent to the symmetry P4′.

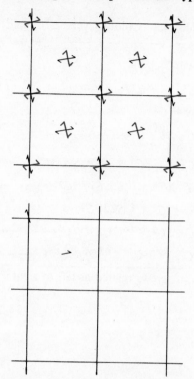

Fig. 14.2—(a) Pattern conforming to the hypothetical space group P_c4' generated from a monatomic motif. (b) An equivalent pattern conforming to P'_4 may be generated from the diatomic motif shown.

14.2 COLOUR SCREW GROUPOIDS

Colour screw operators have the form

$$S\,\mathscr{C}^{1/n}C, \quad S\,\mathscr{C}^{2/n}C^2, \dots \qquad S\,\mathscr{C}^{\frac{n-1}{n}}\,C^{n-1}; \quad C^n = I,$$

with powers

$$S\,\mathscr{C}^{1/n}C, \quad \mathscr{C}^{2/n}C^2, \dots \qquad S^n\,\mathscr{C}^{n/n}\,C^n \text{ i.e. } S^n\mathscr{C}, \text{ etc.,}$$

showing that n must be even for admissible operators. Accordingly, by comparison with the ordinary cyclic screw groupoids (section 10.2), we construct the corresponding colour groupoids

$$\{n'_1\}, \quad \{n'_2\}, \dots \{n'_{n-1}\}; \qquad n = 2, 4, 6 \tag{2}$$

where $\{n'_1\} \Leftrightarrow \{n_1\}$, etc. Reduced versions of all these are included in Table 14.3 and exhibited geometrically in Fig. 14.3.

Table 14.3—Colour Cyclic Groupoids*

$$\{2_1'\} = I, S\mathscr{C}^{1/2} C$$

$$\{4_1'\} = I, S\mathscr{C}^{1/4}C, \mathscr{C}^{1/2}C^2, S\mathscr{C}^{3/4}C^3 \leftarrow$$
$$\{4_2'\} = I, S\mathscr{C}^{1/2}C, C^2, S\mathscr{C}^{1/2}C^3$$
$$\{4_3'\} = I, S\mathscr{C}^{3/4}C, \mathscr{C}^{1/2}C^2, S\mathscr{C}^{1/4}C^3 \leftarrow$$

$$\{6_1'\} = I, S\mathscr{C}^{1/6}C, \mathscr{C}^{1/3}C^2, S\mathscr{C}^{1/2}C^3, \mathscr{C}^{2/3}C^4, S\mathscr{C}^{5/6}C^5 \leftarrow$$
$$\{6_2'\} = I, S\mathscr{C}^{1/3}C, \mathscr{C}^{2/3}C^2, SC^3, \mathscr{C}^{1/3}C^4, S\mathscr{C}^{2/3}C^5 \leftarrow$$
$$\{6_3'\} = I, S\mathscr{C}^{1/2}C, C^2, S\mathscr{C}^{1/2}C^3, C^4, S\mathscr{C}^{1/2}C^5$$
$$\{6_4'\} = I, S\mathscr{C}^{2/3}C, \mathscr{C}^{1/3}C^2, SC^3, \mathscr{C}^{2/3}C^4, S\mathscr{C}^{1/3}C^5 \leftarrow$$
$$\{6_5'\} = I, S\mathscr{C}^{5/6}C, \mathscr{C}^{2/3}C^2, S\mathscr{C}^{1/2}C^3, \mathscr{C}^{1/3}C^4, S\mathscr{C}^{1/6}C^5 \leftarrow$$

*Arrows connect colour screws related as left to right.

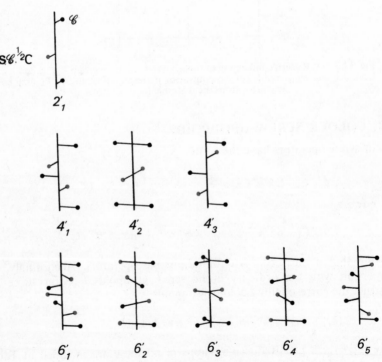

Fig. 14.3—Colour screw groupoids.

Table 14.4—Colour Screw Groupoids

$$\left. \begin{array}{l} \{n_1'22'\} = \{n_1'\} + A^{1/2}\{n_1'\} \\ \{n_1'2'2\} = \{n_1'\} + SA^{1/2}\{n_1'\} \\ \{n_12'2'\} = \{n_1\} + SA^{1/2}\{n_1\} \end{array} \right\}; \; n = 2, 4, 6$$

and similarly for $n_2, n_3, \ldots$ as appropriate.

$$\{2'2_12_1'\} = \{2'\} + \mathscr{A}^{1/2} \, \mathscr{B}^{1/2} \, A\{2'\}$$
$$\{2'2_12_1\} = \{2'\} + S\mathscr{A}^{1/2} \, \mathscr{B}^{1/2} \, A\{2'\}$$
$$\{22_12_1'\} = \{2\} + S\mathscr{A}^{1/2} \, \mathscr{B}^{1/2} \, A\{2\}$$

$$\{4'2_12'\} = \{4'\} + \mathscr{A}^{1/2} \, \mathscr{B}^{1/2} \, A\{4'\}$$
$$\{4'2_12\} = \{4'\} + S\mathscr{A}^{1/2} \, \mathscr{B}^{1/2} \, A\{4'\}$$
$$\{42_12'\} = \{4\} + S\mathscr{A}^{1/2} \, \mathscr{B}^{1/2} \, A\{4\}$$

$$\{2_1'2_12_1'\} = \{2_1'\} + \mathscr{A}^{1/2} \, \mathscr{B}^{1/2} \, A\{2_1'\}$$
$$\{2_1'2_12_1\} = \{2_1'\} + S\mathscr{A}^{1/2} \, \mathscr{B}^{1/2} \, A\{2_1'\}$$
$$\{2_12_12_1'\} = \{2_1\} + S\mathscr{A}^{1/2} \, \mathscr{B}^{1/2} \, A\{2_1\}$$

$$\{4_1'2_12'\} = \{4_1'\} + \mathscr{A}^{1/2} \, \mathscr{B}^{1/2} \, A\{4_1'\}$$
$$\{4_1'2_12\} = \{4_1'\} + S\mathscr{A}^{1/2} \, \mathscr{B}^{1/2} \, A\{4_1'\}$$
$$\{4_12_12'\} = \{4_1\} + S\mathscr{A}^{1/2} \, \mathscr{B}^{1/2} \, A\{4_1\}$$

$$\{4_1'32'\} = \{2_13\} + S\mathscr{A}^{1/4} \, \mathscr{B}^{1/4} \, \mathscr{C}^{-1/4} \, (100)^{1/4}\{2_13\}$$
$$\{4_232'\} = \{23\} + S\mathscr{A}^{1/2} \, \mathscr{B}^{1/2} \, \mathscr{C}^{1/2} \, (100)^{1/2}\{23\}$$
$$\{4_332'\} = \{2_13\} + S\mathscr{A}^{3/4} \, \mathscr{B}^{3/4} \, \mathscr{C}^{-3/4} \, (100)^{3/4}\{2_13\}$$

Corresponding with the dihedral groupoids $\{n_122\}$, etc., of section 11.2, we construct the three distinct types of colour groupoid:

$$\left. \begin{array}{lll} \text{(i)} & \{n_1'22'\}; & n = 2, 4, 6 \\ \text{(ii)} & \{n_1'2'2\}; & n = 2, 4, 6 \\ \text{(iii)} & \{n_12'2'\}; & n = 2, 4 \end{array} \right\} \Leftrightarrow \{n_122\}, \tag{3}$$

built up as shown in Table 14.4. Similarly we construct the colour groupoids

$$\{2'2_12_1'\} \equiv \{2'2_1'2_1\}, \quad \{22_12_1'\} \Leftrightarrow \{22_12_1\} \tag{4}$$

$$\{2_1'2_12_1'\} \equiv \{2'2_1'2_1\} \equiv \{2_12_12_1'\} \Leftrightarrow \{2_12_12_1\}. \tag{5}$$

Further groupoids following this pattern are

$$\{4'2_12'\}, \quad \{4'2_1'2\}, \quad \{42_1'2'\} \Leftrightarrow \{42_12\} \tag{6}$$

$$\{4_1'2_12'\}, \quad \{4_1'2_1'2\}, \quad \{4_12_1'2'\} \Leftrightarrow \{4_12_12\} \tag{7}$$

etc., and

$$\{3_12'1\} \Leftrightarrow \{3_121\}, \quad \{3_11'2'\} \Leftrightarrow \{3_112\} \tag{8}$$

etc. No colour groupoids exist for $n = 6$ other than those included in (3) above.

No colour groupoid can be built from $\{2_13\}$. However, corresponding with $\{4_132\}$, we readily obtain

$$\{4_1'32'\}, \quad \{4_2'32'\}, \quad \{4_3'32'\} \Leftrightarrow \{4_132\} \tag{9}$$

as shown in Table 14.4.

14.3　COLOUR GLIDE GROUPOIDS: MONOCLINIC AND ORTHORHOMBIC SYSTEMS

Colour glide operators have the form

$$S \, \mathscr{A}^{1/2} (001)_m, \quad S \, \mathscr{B}^{1/2} (001)_m, \quad S \, \mathscr{A}^{1/2} \, \mathscr{B}^{1/2} (001)_m, \text{ etc.}, \tag{10}$$

with powers

$$\mathscr{A}, \mathscr{B}, \mathscr{A}\mathscr{B}, \text{ etc.}$$

Accordingly, by comparison with the ordinary glide groupoids (Chapter 12) we construct the corresponding colour groupoids

$$\{a'\} = \{I, S \, \mathscr{A}^{1/2} (001)_m\} \Leftrightarrow \{a\}$$

$$\{b'\} = \{I, S \, \mathscr{B}^{1/2} (001)_m\} \Leftrightarrow \{b\}$$

$$\{g'\} = \{I, S \, \mathscr{A}^{1/2} \, \mathscr{B}^{1/2} (001)_m\} \Leftrightarrow \{g\}$$

which are all geometrically equivalent.

Corresponding with the Abelian groupoid

$$\left\{\frac{2}{a}\right\} = \{2\} + \mathscr{A}^{1/2} (001)_m \{2\},$$

we may construct three distinct colour groupoids

$$
\left.\begin{aligned}
\left\{\frac{2'}{a}\right\} &= \{2'\} + \mathscr{A}^{1/2}\,(001)_m\{2'\} \\[2mm]
\left\{\frac{2}{a'}\right\} &= \{2\} + S\,\mathscr{A}^{1/2}\,(001)_m\{2\} \\[2mm]
\left\{\frac{2'}{a'}\right\} &= \{2'\} + S\,\mathscr{A}^{1/2}\,(001)_m\{2'\}
\end{aligned}\right\} \Leftrightarrow \left\{\frac{2}{a}\right\}.
\tag{12}
$$

Similarly, from

$$
\left\{\frac{2_1}{m}\right\} = \{2_1\} + (001)_m\{2_1\}
$$

we construct

$$
\left.\begin{aligned}
\left\{\frac{2_1'}{m}\right\} &= \{2_1'\} + (001)_m\{2_1'\} \\[2mm]
\left\{\frac{2_1}{m'}\right\} &= \{2_1\} + S\,(001)_m\{2_1\} \\[2mm]
\left\{\frac{2_1'}{m'}\right\} &= \{2_1'\} + S\,(001)_m\{2_1'\}
\end{aligned}\right\} \Leftrightarrow \left\{\frac{2_1}{m}\right\},
\tag{13}
$$

and from

$$
\left\{\frac{2_1}{a}\right\} = \{2_1\} + \mathscr{A}^{1/2}\,(001)_m\{2_1\}
$$

we construct

$$
\left.\begin{aligned}
\left\{\frac{2_1'}{a}\right\} &= \{2_1'\} + \mathscr{A}^{1/2}\,(001)_m\{2_1'\} \\[2mm]
\left\{\frac{2_1}{a'}\right\} &= \{2_1\} + S\,\mathscr{A}^{1/2}\,(001)_m\{2_1\} \\[2mm]
\left\{\frac{2_1'}{a'}\right\} &= \{2_1'\} + S\,\mathscr{A}^{1/2}\,(001)_m\{2_1'\}
\end{aligned}\right\} \Leftrightarrow \left\{\frac{2_1}{a}\right\}.
\tag{14}
$$

There are alternative constructions in terms of J, e.g.

$$\left.\begin{array}{l}\left\{\dfrac{2_1'}{m}\right\}=\{2_1'\}+SJ\{2_1'\}\\[2mm]\left\{\dfrac{2_1}{m'}\right\}=\{2_1\}+SJ\{2_1\}\\[2mm]\left\{\dfrac{2_1'}{m'}\right\}=\{2_1'\}+J\{2_1'\}\end{array}\right\}\Leftrightarrow\left\{\dfrac{2_1}{m}\right\},\tag{15}$$

on bearing in mind

$$\left.\begin{array}{l}J\{2'\}=S(001)_m\{2'\},\quad SJ\{2'\}=S^2(001)_m\{2'\}=(001)_m\{2'\}\\J\{2\}=(001)_m\{2\},\ \text{i.e.}\ SJ\{2\}=S(001)_m\{2\}\end{array}\right\}.\tag{16}$$

Colour dihedral glide groupoids may be constructed on similar lines. Thus, starting with

$$\{2cc\}=\{2\}+\mathscr{C}^{1/2}(010)_m\{2\}=\{I,(001)^{1/2},\ \mathscr{C}^{1/2}(010)_m,\ \mathscr{C}^{1/2}(100)_m\},$$

we immediately obtain

$$\{2'cc'\}=\{2'\}+\mathscr{C}^{1/2}(010)_m\{2'\}=\{I,S(001)^{1/2},\ \mathscr{C}^{1/2}(010)_m,\ S\mathscr{C}^{1/2}(100)_m\}$$
$$\{2'c'c\}=\{2'\}+S\mathscr{C}^{1/2}(010)_m\{2'\}=\{I,S(001)^{1/2},\ S\mathscr{C}^{1/2}(010)_m,\ \mathscr{C}^{1/2}(100)_m\}$$
$$\{2c'c'\}=\{2\}+S\mathscr{C}^{1/2}(010)_m\{2\}=\{I,(001)^{1/2},\ S\mathscr{C}^{1/2}(010)_m,\ S\mathscr{C}^{1/2}(100)_m\}$$

all of which have the property $\Leftrightarrow\{2cc\}$. $\hspace{2cm}$ (17)

Of course no geometrical or physical distinction can be drawn between $\{2'cc'\}$ and $\{2'c'c\}$, since each signifies an ordinary axial glide plane combined with a colour axial glide plane (the combination of these symmetries automatically implies a colour 2-fold axis defined by the line of intersection). So proceeding, we obtain

$$\{2'am'\},\quad\{2'a'm\},\quad\{2a'm'\}\Leftrightarrow\{2am\}\tag{18}$$
$$\{2'gc'\},\quad\{2'g'c\},\quad\{2g'c'\}\Leftrightarrow\{2gc\}\tag{19}$$
$$\{2'ab'\},\quad\{2'a'b\},\quad\{2a'b'\}\Leftrightarrow\{2ab\}\tag{20}$$
$$\{2'gg'\},\quad\{2'g'g\},\quad\{2g'g'\}\Leftrightarrow\{2gg\}\tag{21}$$

Again, no geometrical or physical distinction can be drawn between $\{2'ab'\}$, $\{2'a'b\}$ in (20), and between $\{2'gg'\}$, $\{2'g'g\}$ in (21).' Replacing 2 by 2_1 in accordance with (34), Chapter 12 and following the above pattern, we obtain

$$\{2_1'mc'\},\quad\{2_1'm'c\},\quad\{2_1m'c'\}\Leftrightarrow\{2_1mc\}\tag{22}$$
$$\{2_1'ac'\},\quad\{2_1'a'c\},\quad\{2_1a'c'\}\Leftrightarrow\{2_1ac\}\tag{23}$$

$$\{2_1'gm'\}, \quad \{2_1'g'm\}, \quad \{2_1g'm'\} \Leftrightarrow \{2_1gm\} \tag{24}$$

$$\{2_1'ag'\}, \quad \{2_1'a'g\}, \quad \{2_1a'g'\} \Leftrightarrow \{2_1ag\}. \tag{25}$$

The more complex orthorhombic colour groupoids may be listed as follows:

$$\left.\begin{array}{l}\left\{\dfrac{2}{g'}\dfrac{2}{g'}\dfrac{2}{g'}\right\} = \{222\} + S\mathscr{A}^{1/2}\,\mathscr{B}^{1/2}\,\mathscr{C}^{1/2}\,J\{222\} \\[2ex] \left\{\dfrac{2}{g}\dfrac{2'}{g'}\dfrac{2'}{g'}\right\} = \{22'2'\} + \mathscr{A}^{1/2}\,\mathscr{B}^{1/2}\,\mathscr{C}^{1/2}\,J\{22'2'\} \\[2ex] \left\{\dfrac{2}{g'}\dfrac{2'}{g}\dfrac{2'}{g}\right\} = \{22'2'\} + S\mathscr{A}^{1/2}\mathscr{B}^{1/2}\,\mathscr{C}^{1/2}\,J\{22'2'\}\end{array}\right\} \Leftrightarrow \left\{\dfrac{2}{g}\dfrac{2}{g}\dfrac{2}{g}\right\}, \tag{26}$$

$$\left.\begin{array}{l}\left\{\dfrac{2}{m'}\dfrac{2}{c'}\dfrac{2}{c'}\right\} = \{222\} + S\mathscr{C}^{1/2}\,J\{222\} \\[2ex] \left\{\dfrac{2}{m}\dfrac{2'}{c'}\dfrac{2'}{c'}\right\} = \{22'2'\} + \mathscr{C}^{1/2}\,J\{22'2'\} \\[2ex] \left\{\dfrac{2}{m'}\dfrac{2'}{c}\dfrac{2'}{c}\right\} = \{22'2'\} + S\mathscr{C}^{1/2}\,J\{22'2'\} \\[2ex] \left\{\dfrac{2'}{m'}\dfrac{2}{c}\dfrac{2'}{c'}\right\} = \{2'22'\} + \mathscr{C}^{1/2}\,J\{2'22'\} \\[2ex] \left\{\dfrac{2'}{m}\dfrac{2}{c'}\dfrac{2'}{c}\right\} = \{2'22'\} + S\mathscr{C}^{1/2}\,J\{2'22'\}\end{array}\right\} \Leftrightarrow \left\{\dfrac{2}{m}\dfrac{2}{c}\dfrac{2}{c}\right\}, \tag{27}$$

and similarly

$$\left\{\dfrac{2}{g'}\dfrac{2}{b'}\dfrac{2}{a'}\right\}, \quad \left\{\dfrac{2}{g}\dfrac{2'}{b'}\dfrac{2'}{a'}\right\}, \quad \left\{\dfrac{2}{g'}\dfrac{2'}{b}\dfrac{2'}{a}\right\}, \quad \left\{\dfrac{2'}{g'}\dfrac{2}{b}\dfrac{2'}{a'}\right\}, \quad \left\{\dfrac{2'}{g}\dfrac{2}{b'}\dfrac{2'}{a}\right\} \Leftrightarrow \left\{\dfrac{2}{g}\dfrac{2}{b}\dfrac{2}{a}\right\}. \tag{28}$$

Also, following the same pattern:

$$\left\{\dfrac{2_1}{m'}\dfrac{2}{m'}\dfrac{2}{c'}\right\}, \quad \left\{\dfrac{2_1}{m}\dfrac{2'}{m'}\dfrac{2'}{c'}\right\}, \quad \left\{\dfrac{2_1}{m'}\dfrac{2'}{m}\dfrac{2'}{c}\right\}, \quad \left\{\dfrac{2_1'}{m'}\dfrac{2}{m}\dfrac{2'}{c'}\right\}, \quad \left\{\dfrac{2_1'}{m}\dfrac{2}{m'}\dfrac{2'}{c}\right\},$$

$$\left\{\dfrac{2_1'}{m'}\dfrac{2'}{m'}\dfrac{2}{c}\right\}, \quad \left\{\dfrac{2_1'}{m}\dfrac{2'}{m}\dfrac{2}{c'}\right\} \Leftrightarrow \left\{\dfrac{2_1}{m}\dfrac{2}{m}\dfrac{2}{c}\right\} \tag{29}$$

where—by contrast with (28)—the permutations $2_1'22'$ must now be distinguished from $2_1'2'2$ so providing seven independent groupoids, and similarly

$$\left\{\frac{2_1}{g'}\frac{2}{g'}\frac{2}{a'}\right\}\Leftrightarrow\left\{\frac{2_1}{g}\frac{2}{g}\frac{2}{a}\right\}, \quad \left\{\frac{2_1}{a'}\frac{2}{m'}\frac{2}{g'}\right\}\Leftrightarrow\left\{\frac{2_1}{a}\frac{2}{m}\frac{2}{g}\right\}$$

$$\left\{\frac{2_1}{a'}\frac{2}{c'}\frac{2}{a'}\right\}\Leftrightarrow\left\{\frac{2_1}{a}\frac{2}{c}\frac{2}{a}\right\}, \text{ etc.} \tag{30}$$

Still following the same pattern:

$$\left\{\frac{2}{m'}\frac{2_1}{b'}\frac{2_1}{a'}\right\}, \quad \left\{\frac{2}{m}\frac{2_1'}{b'}\frac{2_1'}{a'}\right\}, \quad \left\{\frac{2}{m'}\frac{2_1'}{b}\frac{2_1'}{a}\right\}, \quad \left\{\frac{2'}{m'}\frac{2_1}{b}\frac{2_1'}{a'}\right\}, \quad \left\{\frac{2'}{m}\frac{2_1}{b'}\frac{2_1'}{a}\right\}$$

$$\Leftrightarrow\left\{\frac{2}{m}\frac{2_1}{b}\frac{2_1}{a}\right\}, \tag{31}$$

as in (28), and similarly

$$\left\{\frac{2}{g'}\frac{2_1}{c'}\frac{2_1}{c'}\right\}\Leftrightarrow\left\{\frac{2}{g}\frac{2_1}{c}\frac{2_1}{c}\right\}, \quad \left\{\frac{2}{m'}\frac{2_1}{g'}\frac{2_1}{g'}\right\}\Leftrightarrow\left\{\frac{2}{m}\frac{2_1}{g}\frac{2_1}{g}\right\}, \quad \left\{\frac{2'}{g'}\frac{2_1}{m'}\frac{2_1}{m'}\right\}$$

$$\Leftrightarrow\left\{\frac{2}{g}\frac{2_1}{m}\frac{2_1}{m}\right\}, \text{ etc.}, \tag{32}$$

as in (27), whilst

$$\left\{\frac{2}{a'}\frac{2_1}{b}\frac{2_1}{m}\right\}\Leftrightarrow\left\{\frac{2}{a}\frac{2_1}{b}\frac{2_1}{m}\right\}, \quad \left\{\frac{2}{a'}\frac{2_1}{g'}\frac{2_1}{c'}\right\}\Leftrightarrow\left\{\frac{2}{a}\frac{2_1}{g}\frac{2_1}{c}\right\}, \text{ etc.}, \tag{33}$$

as in (29). Finally:

$$\left\{\frac{2_1}{b'}\frac{2_1}{c'}\frac{2_1}{a'}\right\}, \quad \left\{\frac{2_1'}{b'}\frac{2_1}{c}\frac{2_1'}{a'}\right\}, \quad \left\{\frac{2_1'}{b}\frac{2_1}{c'}\frac{2_1'}{a}\right\}\Leftrightarrow\left\{\frac{2_1}{b}\frac{2_1}{c}\frac{2_1}{a}\right\} \tag{34}$$

$$\left\{\frac{2_1}{g'}\frac{2_1}{m'}\frac{2_1}{a'}\right\}, \quad \left\{\frac{2_1}{g}\frac{2_1'}{m'}\frac{2_1'}{a'}\right\}, \quad \left\{\frac{2_1}{g'}\frac{2_1'}{m}\frac{2_1'}{a}\right\}, \quad \left\{\frac{2_1'}{g'}\frac{2_1}{m}\frac{2_1'}{a'}\right\}, \quad \left\{\frac{2_1'}{g}\frac{2_1}{m'}\frac{2_1'}{a}\right\}$$

$$\left\{\frac{2_1'}{g'}\frac{2_1'}{m'}\frac{2_1}{a}\right\}, \quad \left\{\frac{2_1'}{g}\frac{2_1'}{m}\frac{2_1}{a'}\right\}\Leftrightarrow\left\{\frac{2_1}{g}\frac{2_1}{m}\frac{2}{a}\right\}. \tag{35}$$

14.4 COLOUR GLIDE GROUPOIDS: TETRAGONAL SYSTEM

Following the pattern of (12)–(14) above, we readily obtain

$$\left\{\frac{4'}{g}\right\}, \quad \left\{\frac{4}{g'}\right\}, \quad \left\{\frac{4'}{g'}\right\}\Leftrightarrow\left\{\frac{4}{g}\right\} \tag{36}$$

$$\left\{\frac{4_2'}{m}\right\}, \quad \left\{\frac{4_2}{m'}\right\}, \quad \left\{\frac{4_2'}{m'}\right\}\Leftrightarrow\left\{\frac{4_2}{m}\right\} \tag{37}$$

$$\left\{\frac{4_2'}{g}\right\}, \quad \left\{\frac{4_2}{g'}\right\}, \quad \left\{\frac{4_2'}{g'}\right\}\Leftrightarrow\left\{\frac{4_2}{g}\right\} \tag{38}$$

on bearing in mind that the possibilities $\left\{\frac{4_1}{m}\right\}, \left\{\frac{4_3}{m}\right\}, \left\{\frac{4}{a}\right\}$ are excluded. Also, following (16), we obtain

$$\{4c'c'\}, \quad \{4'cc'\}, \quad \{4'c'c\}\Leftrightarrow\{4cc\} \tag{39}$$

where $\{4'cc'\}, \{4'c'c\}$ are now independent. Similarly

$$\{4'a'm'\}, \quad \{4g'c'\}, \quad \{4_2m'c'\}, \quad \{4_2a'c'\}\Leftrightarrow\{4am\}, \text{ etc.,} \tag{40}$$

and similarly

$$\{\bar{4}c'2'\}, \quad \{\bar{4}a'2'\}, \quad \{\bar{4}g'2'\}, \quad \{\bar{4}2'c'\}, \quad \{\bar{4}2_1'm'\}, \quad \{\bar{4}2_1'c'\}\Leftrightarrow\{\bar{4}c2\}, \text{ etc.} \tag{41}$$

Finally, following (27) and noting that the permutations $4'2'2$ must be distinguished from $4'22'$, we obtain

$$\left\{\frac{4\ 2\ 2}{m'\ c'\ c'}\right\}, \quad \left\{\frac{4\ 2'\ 2'}{m\ c'\ c'}\right\}, \quad \left\{\frac{4\ 2'\ 2'}{m'\ c\ c}\right\}, \quad \left\{\frac{4'\ 2\ 2'}{m'\ c\ c'}\right\}, \quad \left\{\frac{4'\ 2\ 2'}{m\ c'\ c}\right\}, \quad \left\{\frac{4'\ 2'\ 2}{m'\ c'\ c}\right\},$$

$$\left\{\frac{4'\ 2'\ 2}{m\ c\ c'}\right\}\Leftrightarrow\left\{\frac{4\ 2\ 2}{m\ c\ c}\right\}, \tag{42}$$

and similarly

$$\left\{\frac{4\ 2\ 2}{g'\ b'\ m'}\right\}, \quad \left\{\frac{4\ 2\ 2}{g'\ g'\ c'}\right\}; \quad \left\{\frac{4\ 2_1\ 2}{m'\ b'\ m'}\right\}, \quad \left\{\frac{4\ 2_1\ 2}{m'\ g'\ c'}\right\}, \quad \left\{\frac{4\ 2_1\ 2}{g'\ m'\ m'}\right\}, \quad \left\{\frac{4\ 2_1\ 2}{g'\ c'\ c'}\right\};$$

$$\left\{\frac{4_2\ 2\ 2}{m'\ m'\ c'}\right\}, \quad \left\{\frac{4_2\ 2\ 2}{m'\ c'\ m'}\right\}, \quad \left\{\frac{4_2\ 2\ 2}{g'\ b'\ c'}\right\}, \quad \left\{\frac{4_2\ 2\ 2}{g'\ g'\ m'}\right\};$$

$$\left\{\frac{4_2\ 2_1\ 2}{m'\ b'\ c'}\right\}, \quad \left\{\frac{4_2\ 2_1\ 2}{m'\ g'\ m'}\right\}, \quad \left\{\frac{4_2\ 2_1\ 2}{g'\ m'\ c'}\right\}, \quad \left\{\frac{4_2\ 2_1\ 2}{m'\ c'\ m'}\right\}\Leftrightarrow\left\{\frac{4\ 2\ 2}{g\ b\ m}\right\}, \text{ etc.} \tag{43}$$

14.5 COLOUR GLIDE GROUPOIDS: HEXAGONAL SYSTEM

These may be listed as follows:

$$\left\{\frac{6_3'}{m}\right\}, \quad \left\{\frac{6_3}{m'}\right\}, \quad \left\{\frac{6_3'}{m'}\right\}\Leftrightarrow\left\{\frac{6_3}{m}\right\}, \tag{44}$$

$$\{6c'c'\}, \quad \{6'cc'\}, \quad \{6'c'c\} \Leftrightarrow \{6cc\} \tag{45}$$

$$\{6_3m'c'\}, \quad \{6'_3mc'\}, \quad \{6'_3m'c\} \Leftrightarrow \{6_3mc\} \tag{46}$$

$$\{6_3c'm'\}, \quad \{6'_3cm'\}, \quad \{6'_3c'm\} \Leftrightarrow \{6_3cm\} \tag{47}$$

$$\{3c'1\}, \quad \{31c'\} \Leftrightarrow \{3c1\}, \quad \{31c\}. \tag{48}$$

Also, following the pattern of (42) above:

$$\left\{\frac{6}{m'}\frac{2}{c'}\frac{2}{c'}\right\}, \quad \left\{\frac{6_3}{m'}\frac{2}{m'}\frac{2}{c'}\right\}, \quad \left\{\frac{6_3}{m'}\frac{2}{c'}\frac{2}{m'}\right\} \Leftrightarrow \left\{\frac{6}{m}\frac{2}{c}\frac{2}{c}\right\}, \text{ etc.} \tag{49}$$

Finally,

$$\left. \begin{array}{ccc} \left\{\dfrac{3}{m}c'2'\right\}, & \left\{\dfrac{3}{m'}c'2\right\}, & \left\{\dfrac{3}{m'}c2'\right\} \Leftrightarrow \left\{\dfrac{3}{m}c2\right\} \\[2ex] \left\{\dfrac{3}{m}2'c'\right\} & \left\{\dfrac{3}{m'}2'c\right\}, & \left\{\dfrac{3}{m'}2c'\right\} \Leftrightarrow \left\{\dfrac{3}{m}2c\right\} \end{array} \right\}. \tag{50}$$

14.6   COLOUR GLIDE GROUPOIDS: CUBIC SYSTEM

These may be constructed as follows:

$$\left\{\frac{2}{g'}\bar{3}'\right\} = \{23\} + S\mathscr{A}^{1/2}\mathscr{B}^{1/2}\mathscr{C}^{1/2}J\{23\} \tag{51}$$

$$\left\{\frac{2_1}{a'}\bar{3}'\right\} = \{2_13\} + SJ\{2_13\} \tag{52}$$

$$\{\bar{4}'3g'\} = \{23\} + S\mathscr{A}^{1/2}\mathscr{B}^{1/2}\mathscr{C}^{1/2}J(100)^{1/4}\{23\}. \tag{53}$$

Also

$$\left\{\frac{4}{g'}\bar{3}'\frac{2}{g'}\right\} = \{432\} + S\mathscr{A}^{1/2}\mathscr{B}^{1/2}\mathscr{C}^{1/2}J\{432\} \tag{54}$$

$$\left\{\frac{4'}{g}\bar{3}\frac{2'}{g'}\right\} = \{4'32'\} + \mathscr{A}^{1/2}\mathscr{B}^{1/2}\mathscr{C}^{1/2}J\{4'32'\} \tag{55}$$

$$\left\{\frac{4'}{g'}\bar{3}'\frac{2'}{g}\right\} = \{4'32'\} + S\mathscr{A}^{1/2}\mathscr{B}^{1/2}\mathscr{C}^{1/2}J\{4'32'\}. \tag{56}$$

Similarly

$$\left\{\frac{4_2}{m'}\bar{3}'\frac{2}{g'}\right\}, \quad \left\{\frac{4'_2}{m}\bar{3}\frac{2'}{g'}\right\}, \quad \left\{\frac{4'_2}{m'}\bar{3}'\frac{2'}{g}\right\} \Leftrightarrow \left\{\frac{4_2}{m}\bar{3}\frac{2}{g}\right\} \tag{57}$$

$$\left\{\frac{4_2}{\cdot g'}\,\bar{3}\,\frac{2}{m'}\right\},\quad \left\{\frac{4_2'}{g}\,\bar{3}\,\frac{2'}{m'}\right\},\quad \left\{\frac{4_2'}{g'}\,\bar{3}\,\frac{2'}{m}\right\}\leftrightarrow\left\{\frac{4_2}{g}\,\bar{3}\,\frac{2}{m}\right\}. \tag{58}$$

14.7 SUMMARY

Type I colour space groups are of the form $\{G'\}$ or $\{\Gamma'\}\otimes\{\mathcal{T}\}\otimes\{\mathcal{I}\}$, $\otimes\{\mathcal{F}\}$, $\otimes\{\mathcal{E}\}$, or $\otimes\{\mathcal{R}\}$ as the case may be. All those of the form $\{G'\}\otimes\{\mathcal{T}\}$, $\otimes\{\mathcal{I}\}$, etc., have been covered in Tables 14.1, 14.1a e.g. $C22'2'$, $C2'2'2$. To construct those of the form $\{\Gamma'\}\otimes\{\mathcal{T}\}$, we simply run through every $\{\Gamma'\}$ listed in

Chapter 14 so providing $P\dfrac{2'}{a}$, $P\dfrac{2}{a'}$, $P\dfrac{2'}{a'}$ etc. The independent colour groups of the form $\{\Gamma'\}\otimes\{\mathcal{I}\}$, etc., may be readily constructed by reference to Tables 13.1, 13.2, e.g. $I\dfrac{4_1}{a'}$, $I\dfrac{4_1}{a}$, $I\dfrac{4_1}{a'}$. A breakdown amongst the different crystal systems appears in Table 14.5.

Type II colour space groups are of the form $\{G\}\otimes\{\mathcal{I}'\}$ or $\{\Gamma\}\otimes\{\mathcal{T}'\}$, where $\{\mathcal{T}'\}$ denotes any of the colour translation groups listed in Table 8.1. All those of the form $\{G\}\otimes\{\mathcal{T}'\}$ have been covered in Table 14.2, e.g. R_I3, $F_c\,2mm$. To construct those of the form $\{\Gamma\}\times\{\mathcal{T}'\}$, we must first divide $\{\mathcal{T}'\}$ into two categories corresponding respectively with the unit cells:

(i) P_s, P_a, P_c, P_C, P_I, R_I

Table 14.5—The Type I Colour Space Groups

	PB	CB	PN	CN	D	
triclinic	1	—	—	—	—	1
monoclinic	5	5	11	4	—	25
orthorhombic	6	24	114	64	5	213
tetragonal	24	24	177	17	26	268
cubic	6	12	15	7	11	51
hexagonal	35	6*	71	4*	—	116
	77	71	388	96	42 =	674

PB: primitive Bravais (i.e. symmorphic)   PN: primitive non-Bravais
CB: centred Bravais   CN: centred non-Bravais

D: diamond glide $+I\dfrac{4_1'}{a}$, $\dfrac{4_1}{a'}$, $\dfrac{4_1'}{a'}$

*alternatively classified as primitive rhombohedral

which are primitive with respect to one colour;

    (ii) I_a, I_c, F_c, C_A, C_c, C_a

which are centred with respect to one colour.

    Category (i) space groups may be readily constructed by associating the groupoids of Tables 10.1, 11.1, 12.1 with P_s, etc., as appropriate e.g. P_I2_13. Category (ii) space groups may be constructed by reference to Tables 13.1, 13.2 e.g. $I_c \dfrac{4_1}{d} \bar{3} \dfrac{2}{m}$. A breakdown amongst the different crystal systems appears in Table 14.6.

    The two-dimensional colour space groups are listed in Appendix 8.

Table 14.6—The Type II Colour Space Groups

	PB	CB	PN	CN	D	
triclinic	2	—	—	—	—	2
monoclinic	9	6	21	4	—	40
orthorhombic	11	16	155	47	2	231
tetragonal	24	8	123	5	6	166
cubic	5	5	10	3	3	26
hexagonal	16	5*	29	2*	—	52
	67	40	338	61	10	= 517

PB: primitive Bravais (i.e. symmorphic)   PN: primitive non-Bravais
CB: centred Bravais   CN: centred non-Bravais

D: diamond glide $+I_c \dfrac{4_1}{a}$

*R_I symmetries

Choice of Co-ordinate System

The principal symmetry elements of a point-group provide a natural co-ordinate system for the representation of point-group operators. For instance $\{2/m\}$—see (14), Chapter 12—comprises a 2-fold symmetry axis coupled with a transverse symmetry plane which intersect in a fixed point. It would be natural to identify the symmetry axis as the z-axis of a rectangular Cartesian co-ordinate system, with the symmetry plane as the x, y plane and the fixed point therefore coinciding with $[0, 0, 0]$. Utilising the symbolism of Chapter 3 for rotation operators and of Chapter 12 for reflection operators, we write

$$\left\{\frac{2}{m}\right\} = \{I, (001)^{1/2}, (001)_m, (001)_m, (001)^{1/2}\} \tag{1}$$

where

$$(001)^{1/2} \cdot [x, y, z] = [\bar{x}, \bar{y}, z], \text{ i.e. } (001)^{1/2} = \bar{x}/\bar{y}/z \tag{2}$$

$$(001)_m \cdot [x, y, z] = [x, y, \bar{z}], \text{ i.e. } (001)_m = x/y/\bar{z} \tag{3}$$

$$(001)_m(001)^{1/2} \cdot [x, y, z] = [\bar{x}, \bar{y}, \bar{z}], \text{ i.e. } (001)_m(001)^{1/2} = \bar{x}/\bar{y}/\bar{z} = J. \tag{4}$$

The triads $[\bar{x}, \bar{y}, z]$, etc. define the equivalent points generated by $\{2/m\}$ from $[x, y, z]$, whereas $\bar{x}/\bar{y}/z$, etc., define faithful representations for the operators concerned. Thus, we may supplement (2)–(4) by writing

$$I \cdot [x, y, z] = [x, y, z], \text{ i.e. } I = x/y/z. \tag{5}$$

Operator equations exemplified by the right-hand side of (4) may be proved directly by writing

$$x/y/\bar{z} \cdot \bar{x}/\bar{y}/z = \bar{x}/\bar{y}/\bar{z} \tag{6}$$

where $\bar{x}/\bar{y}/\bar{z}$ appears as the effect of $x/y/\bar{z}$ acting upon $\bar{x}/\bar{y}/z$. Faithful representations of the main point-group operators are listed in Appendix 1b, together with some examples of their use.

The most important groupoid isogonal with $\{2/m\}$—see (17), Chapter 12 —is

$$\left\{\frac{2_1}{a}\right\} = \{I, [001]^{1/2}(001)^{1/2}, [100]^{1/2}(001)_m, [100]^{1/2}[001]^{-1/2}J\}, \tag{7}$$

where

$$[001]^{1/2}(001)^{1/2} = \bar{x}/\bar{y}/(z+\tfrac{1}{2}) \tag{8}$$

$$[100]^{1/2}(001)_m = (x+\tfrac{1}{2})/y/\bar{z} \tag{9}$$

$$(x+\tfrac{1}{2})/y/\bar{z} . \bar{x}/\bar{y}/(z+\tfrac{1}{2}) = (\bar{x}+\tfrac{1}{2})/\bar{y}/(\bar{z}-\tfrac{1}{2}) = [100]^{1/2}[001]^{-1/2}J. \tag{10}$$

By virtue of theorem (19) below, we may write

$$[100]^{1/2}[001]^{-1/2}J = J^* \tag{11}$$

where J^* signifies an inversion through $[\tfrac{1}{4}, 0, \bar{\tfrac{1}{4}}]$. Accordingly $\{2_1/a\}$ contains an inversion centre, and the same applies to every groupoid isogonal with $\{2/m\}$.

Under a translational shift of origin from $[0, 0, 0]$ to $[a, b, c]$, the co-ordinates x, y, z become X, Y, Z where

$$X = x - a, \ Y = y - b, \ Z = z - c. \tag{12}$$

Accordingly, the point-group transformation

$$[x, y, z] \rightarrow G[x, y, z] \tag{13}$$

now appears as

$$[X+a, Y+b, Z+c] \rightarrow G[X+a, Y+b, Z+c]$$
$$\text{i.e. } [X, Y, Z] \rightarrow G[X+a, Y+b, Z+c] - [a, b, c]. \tag{14}$$

For example, if $G = I$ then (14) yields

$$[X, Y, Z] \rightarrow [X+a, Y+b, Z+c] - [a, b, c] = [X, Y, Z]$$

as expected.

If $G = J$, then (14) yields

$$[X, Y, Z] \rightarrow -[X+a, Y+b, Z+c] - [a, b, c]$$
$$= [\bar{X}, \bar{Y}, \bar{Z}] - 2[a, b, c], \tag{15}$$

where $[\bar{X}, \bar{Y}, \bar{Z}]$ signifies an inversion through $[a, b, c]$. Since (15) expresses the same fact as

$$[x, y, z] \to [\bar{x}, \bar{y}, \bar{z}], \tag{16}$$

it follows that

$$[X, Y, Z] \to [\bar{X}, \bar{Y}, \bar{Z}] \tag{17}$$

expresses the same fact as

$$[x, y, z] \to [\bar{x}, \bar{y}, \bar{z}] + 2[a, b, c]. \tag{18}$$

This result proves the theorem stated operationally in (20), Chapter 12, viz. an inversion through $[0, 0, 0]$ accompanied by a translation **t** is equivalent to a direct inversion through

$$\tfrac{1}{2}\mathbf{t} \text{ relative to } [0, 0, 0]. \tag{19}$$

Applying (14) to (8)–(10), we find, respectively, that $[X, Y, Z] \to$

$$[\bar{X} - 2a, \bar{Y} - 2b, Z + \tfrac{1}{2}] \tag{8a}$$
$$[X + \tfrac{1}{2}, Y, \bar{Z} - 2c] \tag{9a}$$
$$[\bar{X} - 2a + \tfrac{1}{2}, \bar{Y} - 2b, \bar{Z} - 2c - \tfrac{1}{2}] \tag{10a}$$

i.e.

$$[\bar{X} - \tfrac{1}{2}, \bar{Y}, Z + \tfrac{1}{2}] \tag{8b}$$
$$[X + \tfrac{1}{2}, Y, \bar{Z} + \tfrac{1}{2}] \tag{9b}$$
$$[\bar{X}, \bar{Y}, \bar{Z}] \tag{10b}$$

on choosing $a = \tfrac{1}{4}$, $b = 0$, $c = -\tfrac{1}{4}$. It will be noted that (10b) is the inverse of $[X, Y, Z]$, and that (9b) is the inverse of (8b), with respect to an origin located at $[\tfrac{1}{4}, 0, -\tfrac{1}{4}]$, in agreement with (11) above.

Starting with the groupoid

$$\left\{\frac{2_1}{c}\right\} = \{I, [010]^{1/2}(010)^{1/2}, [001]^{1/2}(010)_m, [001]^{1/2}[010]^{-1/2}J\}, \tag{20}$$

i.e. $\{2_1/a\}$ in a different orientation, and transferring the origin to $[0, \bar{\tfrac{1}{4}}, \tfrac{1}{4}]$, we obtain the equivalent points given in the International Tables [12], p. 99.

Appendix 1b

Representation of Rotation Operators

For the reader's convenience, we utilise the notation of Bradley & Cracknell [3], where appropriate, as well as that described in Section 1, Chapter 3. Also—see Appendix 1a—we write $x/y/z$, etc., for the faithful representations of point-group operators, which appear as follows:

$$C_{2x} = (100)^{1/2} : x/\bar{y}/\bar{z}, \quad C_{2y} = (010)^{1/2} : \bar{x}/y/\bar{z}, \quad C_{2z} = (001)^{1/2} : \bar{x}/\bar{y}/z$$

$$C_{4x}^+ = (100)^{1/4} : x/\bar{z}/y, \quad C_{4y}^+ = (010)^{1/4} : z/y/\bar{x}, \quad C_{4z}^+ = (001)^{1/4} : \bar{y}/x/z$$

$$C_{4x}^- = (100)^{3/4} : x/z/\bar{y}, \quad C_{4y}^- = (010)^{3/4} : \bar{z}/y/x, \quad C_{4z}^- = (001)^{3/4} : y/\bar{x}/z$$

$$C_{31}^+ = (111)^{1/3} : z/x/y, \quad C_{32}^+ = (\bar{1}\bar{1}1)^{1/3} : \bar{z}/x/\bar{y}, \quad C_{33}^+ = (1\bar{1}\bar{1})^{1/3} : \bar{z}/\bar{x}/y,$$
$$C_{34}^+ = (\bar{1}1\bar{1})^{1/3} : z/\bar{x}/\bar{y}$$

$$C_{31}^- = (111)^{2/3} : y/z/x, \quad C_{32}^- = (\bar{1}\bar{1}1)^{2/3} : y/\bar{z}/\bar{x}, \quad C_{33}^- = (1\bar{1}\bar{1})^{2/3} : \bar{y}/z/\bar{x},$$
$$C_{34}^- = (\bar{1}1\bar{1})^{2/3} : \bar{y}/\bar{z}/x$$

$$(011)^{1/2} : \bar{x}/z/y, \quad (101)^{1/2} : z/\bar{y}/x, \quad (110)^{1/2} : y/x/\bar{z}$$

$$(01\bar{1})^{1/2} : \bar{x}/\bar{z}/\bar{y}, \quad (10\bar{1})^{1/2} : \bar{z}/\bar{y}/\bar{x}, \quad (1\bar{1}0)^{1/2} : \bar{y}/\bar{x}/\bar{z}$$

To prove that $(100)^{1/4}(111)^{1/3} = (101)^{1/2}$—see (6), Chapter 3—we write $(100)^{1/4}(111)^{1/3} . x/y/z = (100)^{1/4} . z/x/y = z/\bar{y}/x = (101)^{1/2} . x/y/z$, etc.

For hexagonal co-ordinates (Fig. 6.14), the relevant operators are

$$C_6^+ = (001)^{1/6} = (x+\bar{y})/x/z, \quad C_2 = (001)^{3/6} : \bar{x}/\bar{y}/z, \quad C_3^- = (001)^{4/6} : (\bar{x}+y)/\bar{x}/z$$

$$C_3^+ = (001)^{2/6} : \bar{y}/(x+\bar{y})/z, \qquad\qquad C_6^- = (001)^{5/6} : y/(\bar{x}+y)/z$$

$$C_{21}'' = (100)^{1/2} : (x+\bar{y})/\bar{y}/\bar{z}, \quad C_{22}'' = (010)^{1/2} : \bar{x}/(\bar{x}+y)/\bar{z}, \quad C_{23}'' = (110)^{1/2} : y/x/\bar{z}$$

$$C_{21}' = (120)^{1/2} : (\bar{x}+y)/y/\bar{z}, \quad C_{22}' = (210)^{1/2} : x/(x+\bar{y})/\bar{z}, \quad C_{23}' = (1\bar{1}0)^{1/2} : \bar{y}/\bar{x}/\bar{z}$$

To prove that $(001)^{1/6}(100)^{1/2} = (100)^{1/2}(001)^{5/6}$—see (31), Chapter 2—we write

$$(001)^{1/6}(100)^{1/2} . x/y/z = (001)^{1/6} . (x+\bar{y})/\bar{y}/\bar{z} = x/(x+\bar{y})/z,$$
$$(100)^{1/2}(001)^{5/6} . x/y/z = (100)^{1/2} . y/(\bar{x}+y)/z = x/(x+\bar{y})/z, \text{ etc.}$$

To prove that

$$[\mathscr{A}^{1/4}\mathscr{B}^{1/4}\mathscr{C}^{-1/4}(100)^{1/4}]^2 = \mathscr{A}^{1/2}\mathscr{B}^{1/2}(100)^{1/2},$$

—see (32), Chapter 11—we note that

$$\mathscr{A}^{1/4}\mathscr{B}^{1/4}\mathscr{C}^{-1/4}(100)^{1/4}.x/y/z = \mathscr{A}^{1/4}\mathscr{B}^{1/4}\mathscr{C}^{-1/4}.x/\bar{z}/y = (x+\tfrac{1}{4})/(\bar{z}+\tfrac{1}{4})/(y-\tfrac{1}{4}),$$
$$\mathscr{A}^{1/4}\mathscr{B}^{1/4}\mathscr{C}^{-1/4}(100)^{1/4}.(x+\tfrac{1}{4})/(\bar{z}+\tfrac{1}{4})/(y-\tfrac{1}{4})$$
$$= (x+\tfrac{1}{4}+\tfrac{1}{4})/(\bar{y}+\tfrac{1}{4}+\tfrac{1}{4})/(\bar{z}+\tfrac{1}{4}-\tfrac{1}{4}) = (x+\tfrac{1}{2})/(\bar{y}+\tfrac{1}{2})/\bar{z} = \mathscr{A}^{1/2}\mathscr{B}^{1/2}(100)^{1/2}.$$

Similarly, to prove that

$$[\mathscr{A}^{1/2}\mathscr{B}^{1/2}\mathscr{C}^{1/2}J(111)^{1/3}]^2 = (111)^{2/3},$$

—see (89), Chapter 12—we note that

$$\mathscr{A}^{1/2}\mathscr{B}^{1/2}\mathscr{C}^{1/2}J(111)^{1/3}.x/y/z = \mathscr{A}^{1/2}\mathscr{B}^{1/2}\mathscr{C}^{1/2}J.z/x/y = (\bar{z}+\tfrac{1}{2})/(\bar{x}+\tfrac{1}{2})/(\bar{y}+\tfrac{1}{2}),$$

from which there follows

$$\mathscr{A}^{1/2}\mathscr{B}^{1/2}\mathscr{C}^{1/2}J(111)^{1/3}.(\bar{z}+\tfrac{1}{2})/(\bar{x}+\tfrac{1}{2})/(\bar{y}+\tfrac{1}{2})$$
$$= (y-\tfrac{1}{2}+\tfrac{1}{2})/(z-\tfrac{1}{2}+\tfrac{1}{2})/(x-\tfrac{1}{2}+\tfrac{1}{2}) = y/z/x = (111)^{2/3}.x/y/z.$$

Our final, easily proved results, are

$$\begin{aligned}
C\mathscr{A}^{\lambda}\mathscr{B}^{\mu}C^{-1} &= \mathscr{A}^{-\lambda}\mathscr{B}^{-\mu}; \quad C = (001)^{1/2}\\
&= \mathscr{A}^{-\mu}\mathscr{B}^{\lambda}; \quad C = (001)^{1/4}\\
&= \mathscr{A}^{-\mu}\mathscr{B}^{\lambda-\mu}; \quad C = (001)^{1/3}\\
&= \mathscr{A}^{\lambda-\mu}\mathscr{B}^{\lambda}; \quad C = (001)^{1/6}.
\end{aligned}$$

Appendix 2

Combination of Crystallographic Axes

A spherical triangle has three sides which are arcs of great circles on the unit sphere, centre O, their lengths a, b, c, being measured by the relevant angles subtended at the centre (Fig. A2.1). These three great circles define three diametral planes intersecting at angles $\hat{A}$, $\hat{B}$, $\hat{C}$ which satisfy the relations

$$\cos c = \frac{\cos \hat{C} + \cos \hat{A} \cos \hat{B}}{\sin \hat{A} \sin \hat{B}} \text{ , etc.} \tag{1}$$

$$\cos \hat{C} = \frac{\cos c - \cos a \cos b}{\sin a \sin b}, \text{ etc.} \tag{2}$$

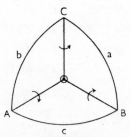

Fig. A2.1

of spherical trigonometry. According to Euler's half-angle construction, a rotation of amount $2\hat{A}$ about the axis AO in the sense indicated, followed by a rotation of $2\hat{B}$ about the axis BO in the sense indicated, is equivalent to a rotation of amount $2\hat{C}$ about the axis OC in the sense indicated. This construction can be understood by reference to Fig. A2.2, which exhibits the triangle reflected into the plane AOB. Regarding C as a test point, it moves into its image position C' under the rotation $2\hat{A}$ and then moves back from C' to C under the rotation $2\hat{B}$; hence, since O remains fixed under these operations, the axis OC remains fixed. Also, reflecting the triangle into the plane BOC (Fig. A2.3), and regarding A as a test point, it moves into its image position A' either (1) by a rotation $2\hat{A}$ followed by a rotation $2\hat{B}$ or (2) by a rotation of $2\hat{C}$ about OC.

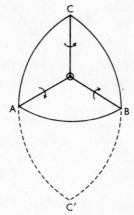

Fig. A2.2

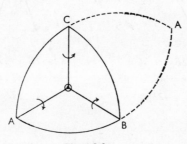

Fig. A2.3

To apply Euler's construction in crystallography, we allow $2\hat{A}$, $2\hat{B}$, $2\hat{C}$ to run through the values $\dfrac{2\pi}{2}, \dfrac{2\pi}{3}, \dfrac{2\pi}{4}, \dfrac{2\pi}{6}$, in all possible combinations, there being $4^3 = 64$ of these at most, substitute into relation (1) above, and admit only those combinations which give

$$0 \leqslant |\cos c| < 1.$$

This procedure leads to the six independent triaxial combinations displayed in Fig. A2.4, corresponding with the pure rotation point groups *222*, *32*,

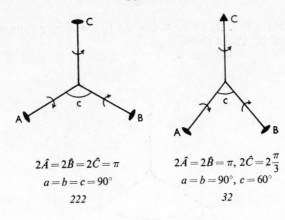

$$2\hat{A} = 2\hat{B} = 2\hat{C} = \pi$$
$$a = b = c = 90°$$
$$222$$

$$2\hat{A} = 2\hat{B} = \pi,\ 2\hat{C} = 2\frac{\pi}{3}$$
$$a = b = 90°,\ c = 60°$$
$$32$$

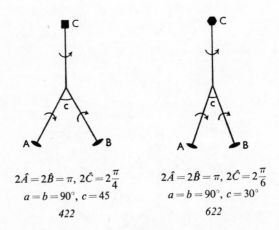

$$2\hat{A} = 2\hat{B} = \pi,\ 2\hat{C} = 2\frac{\pi}{4}$$
$$a = b = 90°,\ c = 45$$
$$422$$

$$2\hat{A} = 2\hat{B} = \pi,\ 2\hat{C} = 2\frac{\pi}{6}$$
$$a = b = 90°,\ c = 30°$$
$$622$$

Fig. A2.4—Combination of an *n*-fold rotation axis with perpendicular 2-fold rotation axis, yielding the symmetries *222, 32, 422, 322*.

422, 622; 23, 432. It will be noted that angles such as $4\pi/3$ are excluded, thus excluding, for instance, the combination

$$2\hat{A} = \frac{2\pi}{4},\quad 2\hat{B} = \frac{2\pi}{4},\quad 2\hat{C} = \frac{4\pi}{3};\quad \cos c = 0,\quad c = 90°,$$

which appertains to the point group *432* as shown in Fig. A2.5.

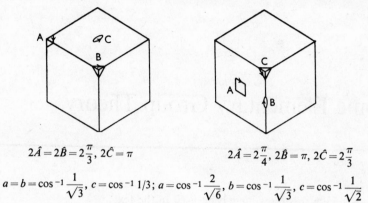

$$2\hat{A}=2\hat{B}=2\frac{\pi}{3},\ 2\hat{C}=\pi$$

$$a=b=\cos^{-1}\frac{1}{\sqrt{3}},\ c=\cos^{-1}1/3;$$

23

$$2\hat{A}=2\frac{\pi}{4},\ 2\hat{B}=\pi,\ 2\hat{C}=2\frac{\pi}{3}$$

$$a=\cos^{-1}\frac{2}{\sqrt{6}},\ b=\cos^{-1}\frac{1}{\sqrt{3}},\ c=\cos^{-1}\frac{1}{\sqrt{2}}$$

432

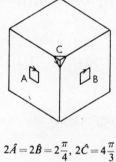

$$2\hat{A}=2\hat{B}=2\frac{\pi}{4},\ 2\hat{C}=4\frac{\pi}{3}$$

$$a=b=\cos^{-1}\frac{1}{\sqrt{3}},\ c=90°$$

432

Fig. A2.5

Appendix 3

Some Elementary Group Theory

The following theorems have been used in the text:

1. If $\{G\}$ is a group, then $\{G\} + J\{G\}$ is a group as follows from $J\{G\} = \{G\}J$.
2. The set
$$\{23\} = \{222\} + (111)^{1/3}\{222\} + (111)^{2/3}\{222\}$$

is a group, as follows from

$$(111)^{1/3}\{222\} = \{222\}(111)^{1/3}, \quad (111)^{2/3}\{222\} = \{222\}(111)^{2/3}.$$

3. The set

$$\{432\} = \{23\} + (100)^{1/4}\{23\}$$

is a group, as follows from

$$(100)^{1/4}\{23\} = \{23\}(100)^{1/4}.$$

4. If $X_1, X_2, \ldots X_r$ are the r independent generators of $\{G\}$, then the maximum number of subgroups of index 2 in $\{G\}$ is 2^{r-1}. This theorem has been proved by B. M. Hurley as follows. If $\{H\}$ is a subgroup of $\{G\}$ with index 2 in $\{G\}$, then $\{H\}$ is normal in $\{G\}$. Hence the factor group $\{G\}/\{H\}$ is cyclic of order 2. Thus $\{H\}$ is the kernel of some homomorphism

$$\theta : \{G\} \rightarrow \{G'\},$$

where $\{G'\}$ is the cyclic group of order 2, with generator $\bar{X}$ say. Each homomorphism is determined by its effect on the generators. Each generator can be mapped either onto the identity I or onto $\bar{X}$, giving 2^r choices. However, the homomorphism that maps every generator onto I does not have as an image the whole of $\{G'\}$. This leaves at most $2^r - 1$ homomorphisms whose image is $\{G'\}$ and each homomorphism determines a different $\{H\}$.

Class Structure of Point Groups

---

1. The operators $\{L_1, L_2, \ldots L_m\}$ of $\{G\}$ belong to the same class if

$$GL_iG^{-1} \subset \{L_1, L_2, \ldots L_m\}; \; i = 1, 2, \ldots m$$

for every member of $\{G\}$. Clearly, in the case of an Abelian group, each member forms its own class.

2. The classes of the dihedral groups (32), Chapter 2, are

$$I; \; C, C^2; \; D, CD, C^2D \qquad\qquad n = 3$$
$$I; \; C^2; \; C, C^3; \; D, C^3D; \; CD, C^2D; \; n = 4$$
$$\left.\begin{array}{l} I; \; C^3; \; C, C^5; \; C^2, C^4 \\ D, C^2D, C^4D; \; CD, C^3D \end{array}\right\} \quad n = 6$$

3. The classes of $\{23\}$ are

$$I; \; (100)^{1/2}, \text{ etc.}; \; (111)^{1/3}, (111)^{2/3}; \text{ etc.}$$

Also, the classes of $\{432\}$ are

$$I; \; (100)^{1/2}, \text{ etc.}; \; (111)^{1/3}, (\bar{1}\bar{1}1)^{1/3}, \text{ etc.}; \; (011)^{1/2}, \text{ etc.};$$
$$(100)^{1/4}, (100)^{3/4}, \text{ etc.}; \; (111)^{2/3}, (\bar{1}\bar{1}1)^{2/3}, \text{ etc.}$$

4. If $\{L_1, L_2, \ldots L_m\}$ forms a class of $\{G\}$, then it also forms a class of $\{G\} + J\{G\}$, as does $J\{L_1, L_2, \ldots L_m\}$.

Packing and Stacking of Lattice Planes

All the parallel lattice planes $hx + ky + lz = 0, \pm 1, \pm 2, \ldots$, have the same arrangement of lattice points, as discussed in Chapter 5. To examine the geometrical relation between a plane and its successor, we refer the lattice to three vectors

$$\mathbf{a}' = l\mathbf{a} - h\mathbf{c}, \quad \mathbf{b}' = l\mathbf{b} - k\mathbf{c}, \quad \mathbf{c}' = p\mathbf{a} + q\mathbf{b} + r\mathbf{c}, \tag{1}$$

where $\mathbf{a}'$, $\mathbf{b}'$ lie in the plane $hx + ky + lz = 0$ and where $\mathbf{c}'$ is a vector perpendicular to these sufficiently defined by

$$\mathbf{c}'.\mathbf{a} = \mathbf{c}'.\mathbf{b} = 0. \tag{2}$$

Hence, any lattice vector $x\mathbf{a} + y\mathbf{b} + z\mathbf{c}$ gets relabelled $x'\mathbf{a}' + y'\mathbf{b}' + z'\mathbf{c}'$ where x', y', z' are not necessarily integers. Eliminating $\mathbf{a}'$, $\mathbf{b}'$, $\mathbf{c}'$ by virtue of (1) from the equation

$$x\mathbf{a} + y\mathbf{b} + z\mathbf{c} = x'\mathbf{a}' + y'\mathbf{b}' + z'\mathbf{c}'$$

yields at once the coordinate transformation

$$\begin{pmatrix} x \\ y \\ z \end{pmatrix} = \begin{pmatrix} l & o & p \\ o & l & q \\ h & k & r \end{pmatrix} \begin{pmatrix} x' \\ y' \\ z' \end{pmatrix}, \tag{3}$$

of which the inverse is

$$x' = \frac{1}{l}\left(x - \frac{pC}{hp + kq + lr} \right),$$

$$y' = \frac{1}{l}\left(y - \frac{pC}{hp + kq + lr} \right),$$

$$z' = \frac{C}{hp + kq + lr}, \tag{4}$$

where $C = hx + ky + lz$. Note that x', y' depend only on the ratio $p:q:r$, in contrast to z'. Distances along the z'-direction are given by

$$z'c' = \frac{C}{hp + kq + lr} \cdot |p\mathbf{a} + q\mathbf{b} + r\mathbf{c}| \tag{5}$$

which again depends only on $p:q:r$. Putting $C = 1$ yields the interplanar spacing d, the relevant vector being

$$\mathbf{d} = \frac{p\mathbf{a} + q\mathbf{b} + r\mathbf{c}}{hp + kq + lr}. \tag{6}$$

Expressions (5) and (6) differ from the corresponding reciprocal lattice formulae (see [1], Chapter 7) in involving $\mathbf{a}.\mathbf{b}, \mathbf{b}.\mathbf{c}, \mathbf{c}.\mathbf{a}$ rather than $\mathbf{a} \times \mathbf{b}, \mathbf{b} \times \mathbf{c}, \mathbf{c} \times \mathbf{a}$.

The most convenient choice of p, q, r as defined by (2) is

$$\begin{pmatrix} p \\ q \\ r \end{pmatrix} = \left\{ \begin{matrix} \frac{bc}{a}(1 - \cos^2 \alpha) & & \\ c(\cos \alpha \cos \beta - \cos \gamma) & \frac{ca}{b}(1 - \cos^2 \beta) & \\ b(\cos \gamma \cos \alpha - \cos \beta) & a(\cos \beta \cos \gamma - \cos \alpha) & \frac{ab}{c}(1 - \cos^2 \gamma) \end{matrix} \right\} \begin{pmatrix} h \\ k \\ l \end{pmatrix} \tag{7}$$

where $\cos \alpha = (\mathbf{b}.\mathbf{c})/bc$, etc. In the case of an orthorhombic cell, where $\alpha = \beta = \gamma = 90°$, we have

$$p = \frac{bc}{a}h, \quad q = \frac{ca}{b}k, \quad r = \frac{ab}{c}l. \tag{8}$$

Putting $C = 1$ in (4) gives the x', y', z' of a lattice point x_1, y_1, z_1 in the plane $hx + ky + lz = 1$, of which the projection on $hx + ky + lz = 0$ is x', y', 0; the join of x', y', 0 to the nearest available lattice point in $hx + ky + lz = 0$ defines the stacking vector $\mathbf{t}$. Each plane is generated from its predecessor by a rigid-body translation $\mathbf{d} + \mathbf{t}$. Generally speaking, it would be necessary to obtain x_1, y_1, z_1 by inspection, but if $l = 1$ an immediate possibility is 0, 0, 1 so that

$$x' = -\frac{p}{hp + kq + lr}, \quad y' = -\frac{q}{hp + kq + lr}.$$

Similar considerations apply if $h = 1$ or $k = 1$. As an illustration of the analysis, we have mapped in Fig. A5.1 the primitive cubic lattice plane $2x + 9y + 5z = 0$, together with the projections $\bigcirc$, $\odot$ of the lattice points

1, 1, $\bar{2}$; 2, 2, $\bar{4}$ lying in $2x + 9y + 5z = 1$, 2 respectively. The coordinates of H relative to $\mathbf{a'}$, $\mathbf{b'}$, $\mathbf{c'}$ are

$$x' = \tfrac{1}{5} \cdot \tfrac{108}{110}, \quad y' = \tfrac{1}{5} \cdot \tfrac{101}{110}, \quad z' = 0$$

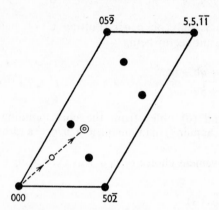

Fig. A5.1 —Stacking of (295) planes in the primitive cubic lattice. The symbols $\bigcirc$, $\circledcirc$ mark respectively the projection of lattice points lying in the two successive planes to the plane mapped, the height of the points being d_{295}, $2d_{295}$ respectively.

and its coordinates relative to $\mathbf{a}$, $\mathbf{b}$, $\mathbf{c}$ are

$$x = \tfrac{108}{110}, \quad y = \tfrac{101}{110}, \quad z = \tfrac{\overline{225}}{110}.$$

Evidently, therefore, the projection of 110 [1, 1, $\bar{2}$] will coincide with a lattice point of $2x + 9y + 5z = 0$, showing that the stacking pattern of primitive cubic (295) planes constitutes a congruence modulo 110. This result can be readily generalised to (hkl) planes on noting from (8) that $p = h$, $q = k$, $r = l$ for a primitive cubic cell.

A lattice referred to a non-primitive cell, e.g. a body-centred cell, may be regarded as consisting of two interpenetrating lattices A and B, the latter being generated from the former by a rigid-body translation $[\tfrac{1}{2}, \tfrac{1}{2}, \tfrac{1}{2}]$. Assuming that our previous analysis applies to the A-lattice, there are two possibilities for B-lattice points in relation to (hkl) planes:

(i) They lie on these planes, the condition being

$$(h + k + l)/2 = \text{integer}.$$

If so, the array of A- and B-points in any such plane constitute a two-dimensional lattice.

(ii) They do not lie on these planes, in which case they necessarily form a set of (hkl) planes falling half-way between the original (hkl) planes, being generated from the latter by a rigid-body translation $\tfrac{1}{2}(\mathbf{d} + \mathbf{t})$.

Composite Operators

A rotation axis is imagined to pass through the origin of coordinates perpendicular to the x, y plane. Utilising complex numbers, this transforms $x + iy$ into $(x + iy)\, e^{i\theta}$ where θ is the angle of rotation. Superposing a translation of amount t parallel to Ox gives (Fig. A6.1).

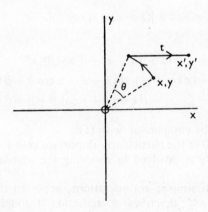

Fig. A6.1

$$x' + iy' = (x + iy)\, e^{i\theta} + t = (x \cos\theta - y \sin\theta + t) + i(x \sin\theta + y \cos\theta). \quad (1)$$

An alternative route for $x + iy \rightarrow x' + iy'$ is by means of a rotation through θ, in the same sense, about a parallel axis passing through $a + i\beta$ where (Fig. A6.2).

$$\alpha = t/2, \quad \beta = \frac{t}{2}\frac{\sin\theta}{1 - \cos\theta} = \frac{t}{2}\cot\frac{\theta}{2}. \quad (2)$$

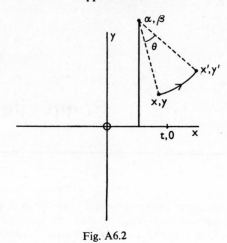

Fig. A6.2

Thus, writing

$$x + iy = \alpha + i\beta + [(x - \alpha) + i(y - \beta)],$$

we have

$$x' + iy' = \alpha + i\beta + [(x - \alpha) + i(y - \beta)]\, e^{i\theta}$$
$$= (x \cos \theta - y \sin \theta + \alpha - \alpha \cos \theta + \beta \sin \theta)$$
$$+ i(x \sin \theta + y \cos \theta + \beta - \beta \cos \theta - \alpha \sin \theta) \quad (3)$$

whence (2) follows by comparison with (1).

Note that $\beta = 0$ in the particularly important case $\theta = \pi$. Also note that no loss of generality is involved in choosing the translation to be parallel to Ox.

Similar, though simpler considerations, apply to theorem (6), section 12.1. Thus, since M describes a reflection through (001), we have $M[x, y, z] = [x, y, \bar{z}]$. Also, since $\mathscr{C}^\nu$ describes a translation of amount νc along [001], it follows that

$$[x', y', z'] = \mathscr{C}^\nu M[x, y, z] = \mathscr{C}^\nu [x, y, \bar{z}] = [x, y, \bar{z} + \nu].$$

An alternative route to $[x', y', z']$ is by a reflection through the plane $z - \nu/2 = 0$. Thus, writing

$$[x, y, z] = [x, y, z - \nu/2] + [0, 0, \nu/2],$$

the reflected point is seen to be

$$[x, y, -z + \nu/2] + [0, 0, \nu/2] = [x, y, \bar{z} + \nu]$$

in agreement with $[x', y', z']$ above.

To prove that $M\mathscr{C}^\nu = \mathscr{C}^{-\nu}M$, we merely note that

$$M\mathscr{C}[x, y, z] = M[x, y, z + \nu] = [x, y, \bar{z} - \nu],$$
$$\mathscr{C}^{-\nu}M[x, y, z] = \mathscr{C}^{-\nu}[x, y, \bar{z}] = [x, y, \bar{z} - \nu].$$

as stated in (7), section 12.1.

Appendix 7

Two-dimensional Space Groups

Reference to Chapter 1 shows that the ten point groups

$$1, 2, 3, 4, 6; \quad m, 2mm, 3m, 4mm, 6mm$$

produce symmetry patterns in a plane, so yielding at once the ten two-dimensional space groups $P1, \ldots$. For $P1$, the plane of the pattern may be any lattice plane. For Pm it is (100) or (010) relative to the monoclinic unit cell, with (001) providing the mirror effect. For all the remaining cases, the plane of the pattern if (001) or (0001), with mirror effects provided by appropriate planes through the principal axis. We note that $P3m$ expands into the two possibilities $P3m1$, $P31m$. Mirror reflection may be replaced by

Table A7.1   Two-dimensional space-group patterns

Crystal system	Lattice type	Symmetry	Plane of pattern	Reflection plane or glide plane
triclinic	P	*1*	any plane	—
monoclinic	P	*2*	(001)	—
	P	*m, a*	(100) or (010)	(001)
	I	*m*	,,	,,
orthorhombic	P	*2mm*	(001)	(100), (010)
		2am, 2ab		
	C	*2mm*	,,	,,
tetragonal	P	*4*	,,	(100), etc.
	P	*4mm, 4ab*	,,	,,
hexagonal	P	*3, 6*	(0001)	—
		3m1	,,	($10\bar{1}1$), etc.
		31m	,,	($12\bar{3}0$), etc.
		6mm	,,	($10\bar{1}1$), ($12\bar{3}0$), etc.

glide reflection for P*m*, P*2mm*, P*4mm* so producing the additional two-dimensional space groups P*a*, P*2am*, P*2ab*, P*4ab*. Finally the rectangle may be centred for P*m*, P*2mm* so yielding the centred two-dimensional space groups A*m*, C*2mm*. All these appear in Table A7.1, except that A*m* is replaced by the equivalent though more symmetrical I*m* (see Table 9.1).

Patterns conforming to the 17 two-dimensional space groups are exhibited in Fig. A7.1 following Bhagavantam and Venkatarayudu [9].

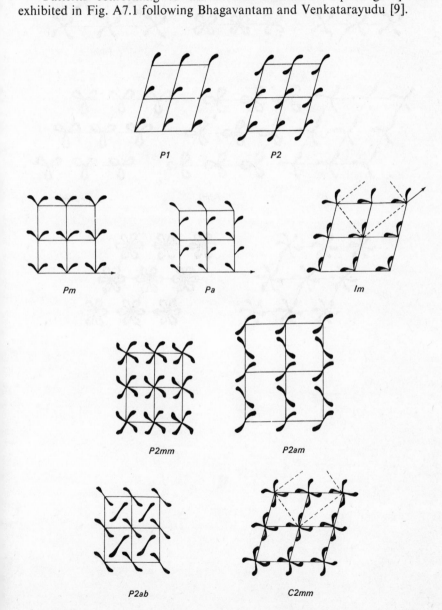

Appendixes

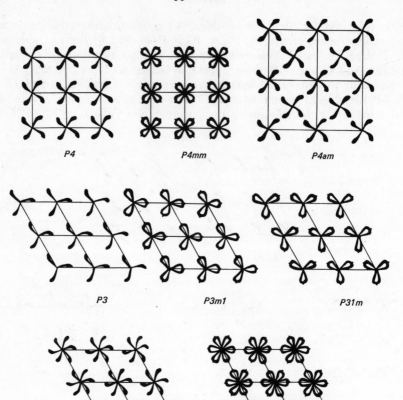

P4

P4mm

P4am

P3

P3m1

P31m

P6

P6mm

Two-dimensional Colour Space Groups

Following section 14.1, the two-dimensional colour space groups fall into two types:

Type I—the direct product of a two-dimensional colour point group or groupoid with an uncoloured two-dimensional lattice.

Type II—the direct product of an ordinary two-dimensional point group or groupoid with a colour two-dimensional lattice.

Starting with Table A7.1 as a reference, we run through Chapter 14 and pick out the colour point groups or groupoids which provide two-dimensional symmetries. These are listed in Table A8.1 against the appropriate uncoloured lattices, so yielding the 26 type I space groups.

Table A8.1 Type I Two-dimensional Space Groups

Crystal system	Lattice type	Plane	Symmetries	Mirror plane
monoclinic	P	(001)	$2'$	—
		(100)	m', a'	(001)
	I	,,	m'	,,
orthorhombic	P	(001)	$2'mm'$, $2m'm'$, $2'ab'$, $2a'b'$ $2'am'$, $2'a'm$, $2a'm'$	(100), (010)
	C	,,	$2'mm'$, $2m'm'$	,,
tetragonal	P*	,,	$4'$ $4'mm'$, $4m'm'$ $4'am'$, $4'a'm$, $4a'm'$	— (100), (110) etc.
hexagonal	P*	(0001)	$6'$	—
		,,	$3m'1$	$(10\bar{1}0)$, etc.
		,,	$31m'$	$(12\bar{3}0)$, etc.
		,,	$6'mm'$, $6m'm'$	$(10\bar{1}0)$, $(12\bar{3}0)$, etc.

*Note the alternative settings P$4'm'm$, P$6'm'm$ listed in Table 14.1a.

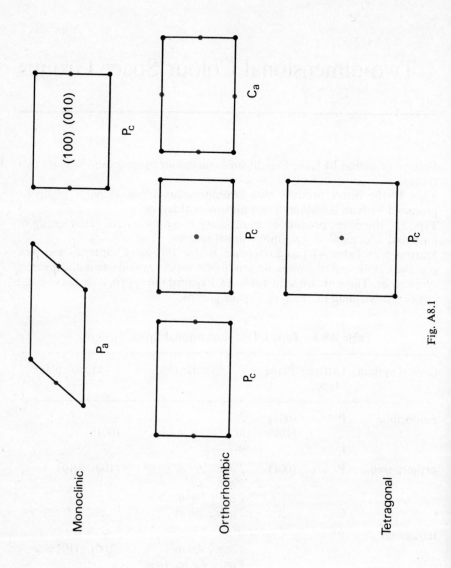

Fig. A8.1

The five two-dimensional colour lattices may be deduced directly or by reference to the colour unit cells displayed in Fig. 8.3. These appear in Fig. A8.1, each labelled by the appropriate unit cell and generating co-ordinate plane. The relevant uncoloured point groups or groupoids may be selected without difficulty, so yielding the 20 type II space groups listed in Table A8.2. Accordingly we obtain the 46 two-dimensional colour space groups.

Table A8.2 Type II Two-dimensional Space Groups

Crystal system	Lattice type	Plane	Symmetries	Mirror plane
monoclinic	P_a	(001)	*1*, *2*	—
	P_c	(100)	*m**, *a**	(001)
orthorhombic	P_c	(100)	*m*, *a*	(001) or (010)
	,,	,,	*2mm*, *2am**, *2ab*	(001), (010)
	P_C	(001)	*m*, *a*	(100) or (010)
	,,	,,	*2mm*, *2am*, *2ab*	(100), (010)
	C_a	,,	*m*, *2mm*	,,
tetragonal	P_C	(001)	*4*	—
	,,	,,	*4mm*, *4am*	(100), (110), etc.

*covered by orthorhombic $P_c(100)$ symmetries *m*, *a* for which the mirror plane may be either (001) or (010) yielding the alternative settings $P_c m1$, $P_c 1m$; $P_c a1$, $P_c 1a$; and the corresponding alternative settings $P_c 2am$, $P_c 2mb$.

Names and Symbols for the Crystal Classes†

System	Class name	International symbol	Short symbol	Schoenflies symbol
triclinic	triclinic-pedial	1		C_1
	triclinic-pinacoidal	$\bar{1}$		C_i
monoclinic	monoclinic-domatic	m		C_g
	monoclinic-sphenoical	2		C_2
	monoclinic-prismatic	$\dfrac{2}{m}$		C_{2h}
orthorhombic	orthorhombic-pyramidal	$2mm$	mm	C_{2v}
	orthorhombic-disphenoidal	222		D_2
	orthorhombic-dipyramidal	$\dfrac{2}{m}\dfrac{2}{m}\dfrac{2}{m}$	mmm	D_{2h}
tetragonal	tetragonal-disphenoidal	$\bar{4}$		S_4
	tetragonal-pyramidal	4		C_4
	tetragonal-dipyramidal	$\dfrac{4}{m}$		C_{4h}
	tetragonal-scalenohedral	$\bar{4}m2$ (or $\bar{4}2m$)		D_{2d}
	ditetragonal-pyramidal	$4mm$		C_{4v}
	tetragonal-trapezohedral	422		D_4
	ditetragonal-dipyramidal	$\dfrac{4}{m}\dfrac{2}{m}\dfrac{2}{m}$	$\dfrac{4}{mmm}$	D_{4h}
trigonal or hexagonal	trigonal-pyramidal	3		C_3
	rhombohedral	$\bar{3}$		C_{3i}
	ditrigonal-pyramidal	$3m1$ (or $31m$)	$3m$	C_{3v}
	trigonal-trapezohedral	321 (or 312)	32	D_3
	ditrigonal-scalenohedral	$\bar{3}\dfrac{2}{m}1$ $\left(\text{or } \bar{3}1\dfrac{2}{m}\right)$	$\bar{3}m$	D_{3d}

Names and Symbols for the Crystal Classes—continued

System	Class name	International symbol	Short symbol	Schoenflies symbol
hexagonal	trigonal-dipyramidal‡	$\dfrac{3}{m} \equiv \bar{6}$		C_{3h}
	ditrigonal-dipyramidal‡	$\dfrac{3}{m}m2 \equiv \bar{6}m2$ $\left(\text{or } \dfrac{3}{m}2m = \bar{6}2m\right)$		D_{3h}
	hexagonal-pyramidal	6		C_6
	hexagonal-dipyramidal	$\dfrac{6}{m}$		C_{6h}
	dihexagonal-pyramidal	$6mm$		C_{6v}
	hexagonal-trapezohedral	622		D_6
	dihexagonal-dipyramidal	$\dfrac{6}{m}\dfrac{2}{m}\dfrac{2}{m}$	$\dfrac{6}{mmm}$	D_{6h}
cubic	tetratoidal	23		T
	diploidal	$\dfrac{2}{m}\bar{3}$	$m3$	T_h
	hextetrahedral	$\bar{4}3m$		T_d
	gyroidal	432		O
	hexoctahedral	$\dfrac{4}{m}\bar{3}\dfrac{2}{m}$	$m3m$	O_h

† This is the term conventionally used by crystallographers for the crystallographic point groups.

‡ the symbol $\bar{6}$ is conventionally used in place of $\dfrac{3}{m}$ to emphasise that the underlying space lattice can only be primitive hexagonal.

References

1. Jaswon, M. A. (1965). *An Introduction to Mathematical Crystallography*. Longmans: London.
2. Rose, M. A. (1975). *The Crystallographic Double Groups: A Fresh Approach*. Ph.D. Thesis. The City University, London.
3. Bradley, C. J., and Cracknell, A. D. (1972). *The Mathematical Theory of Symmetry in Solids*. Clarendon Press: Oxford.
4. Shubnikov, A. V., and Koptsick, V. A. (1974). *Symmetry in Science and Art*. Plenum Press: New York.
5. Phillips, F. C. (1963). *An Introduction to Crystallography*. Longmans: London.
6. Buerger, M. J. (1956). *Elementary Crystallography*. Wiley: New York.
7. Wilson, A. J. C. (1970). *Elements of X-Ray Crystallography*. Addison-Wesley: Mass.
8. Woolfson, M. M. (1970). *An Introduction to X-Ray Crystallography*. Cambridge University Press: London.
9. Bhagavantam, S., and Venkatarayudu, T. (1962). *Theory of Groups and its Application to Physical Problems*. Andhra University Press: India.
10. Lederman, W. (1961). *Introduction to the Theory of Finite Groups*. Oliver & Boyd: Edinburgh.
11. Bravais, A. J., *École Polytech. Paris*, **19** (1850); English translation, Crystallographic Society of America, *Memoir No. 1* (1949). Also *J. École Polytech. Paris*, **20** (1851), 101.
12. International Tables for X-Ray Crystallography. Vol. I (1969). Edited by N. F. Henry, and Kathleen Lonsdale. The Kynoch Press: Birmingham.
13. Hilton, H. (1963). *Mathematical Crystallography*. Dover Edition of 1903 publication.
14. Schwarzenberger, R. L. E. (1980). *N-Dimensional Crystallography*. Pitman: London.
15. Loeb, A. L. (1972). *Color and Symmetry*. Wiley: New York.
16. Jaswon, M. A., and Dove, D. B., *Acta Cryst.* (1955), **8**, 88, and **8**, 806; (1956), **9**, 621; (1957), **10**, 14; (1960), **13**, 232.
17. Jaswon, M. A. (1959). *Studies in Crystal Physics*. Butterworths: London.
18. Janssen, T. (1973). *Crystallographic Groups*. North-Holland: Amsterdam.

19. Macgillavry, C. A. (1965). *Symmetry Aspects of M. C. Escher's Periodic Drawings*. Published for the International Union of Crystallography.
20. Bevis. M., *Acta Cryst.* (1969). A**25,** 370.
21. Nicholas, J. F., *Acta Cryst.* (1970), A**26,**. 470.
22. Gruber, B., *Acta Cryst.* (1970). A**26,** 622; (1973), A**29,** 433.
23. Kolakowski, B. *App. Physics* (1979), **20,** 305.

Index

B

Bevis, M. 67, 187
Bhagavantam, S. 11, 186
Bradley, C. J. 10, 164, 186
Bravais, A. J. 11, 104, 186
bravais
  cell 74, 75, 86
  colour cell 96, 97
  lattice 80, 81, 89
  space group 98–101, 143
Buerger, M. J. 10, 12, 186

C

colour
  glide groupoid 152–159
  operator 48
  point group 11, 48, 49, 59, 60
  screw groupoid 149–151
  space group 145–148, 159, 160
  symmetry 15, 48
  translation group 93
  translation operator 90–92
  unit cell 96, 97
Cracknell, A. D. 10, 164, 186
crystal
  class 184, 185
  system 67, 73, 100
  three-dimensional 76
  two-dimensional 61, 70–73

D

degeneracies 47
diamond
  crystal 184, 185
  glide 134, 138, 142
  groupoid 140
  space group 140, 143
  structure 82, 83
direction-ratio 37
Dove, D. B. 186

E

edge-centring 94, 95
end-centring 83, 84
equivalent
  atoms 12, 26, 27, 98, 108, 163
  space groups 11, 134, 135
Euler
  half-angle construction 166–169

F

factor group 100, 108

G

glide
  groupoid 131, 132
  operator 12, 119, 120
  plane 11, 15, 104
Grey group 48
groupoid 11, 12, 105, 107
  choice of origin 12, 162, 163
  theorem 133
  see Diamond
  see Glide
  see Screw
Gruber, B. 67, 187

H

hexagonal
  cell 73–75, 84
  close-packed 76–79
  net 69, 72, 73, 76, 85, 103
Hilton, H. 12, 186
holohedral
  point group 90
  symmetry 74, 76, 86, 100, 101
honeycomb net 79, 85
Hurley, B. M. 49, 110, 170

I

international tables 12, 163
inversion
    centre 17, 18, 23–25, 162
    theorem 121, 163
isogonal 12, 107, 108

J

Janssen, T. 186
Jaswon, M. A. 186, 187

K

Kolakowski, B. 48, 187
Kopstick, V. A. 10, 186

L

lattice
    plane 64–66, 172–174
    point 62–65, 89
    three-dimensional 64, 70
    two-dimensional 61–63, 67, 68
    vector 62–65, 88, 89
lattice cell
    see Unit cell
Lederman, W. 11, 186
Loeb, A. L. 12, 186

M

Macgillavry, C. A. 187
macropscopic 11, 15, 76–79, 86, 104, 108
microscopic 11, 15, 76–79, 104, 108
Miller indices 64, 86
motif
    asymmetric 76, 79
    pattern 11, 70–73, 76–79, 98, 99
    unit 18, 19

N

Nicholas, J. F. 67, 187

O

octahedral
    group 40
    symmetry 44
operator 26
    colour (see Colour)
    composite 110, 111, 119, 120, 175–177
    inversion 28, 44
    reflection 27, 119, 161
    representation 164, 165
    translation 88

P

Phillips, F. C. 10, 186
point group 11, 12, 25–27, 36, 37, 47
    class structure 171
    colour (see Colour)
    subgroup 34, 35, 48
primitive cell 62, 64, 74

R

reflection 12, 24, 26
rhombohedral 75, 84–86
Rose, M. A. 186
rotation 16, 27, 32, 37, 163, 164
    see also Euler
roto-inversion 29, 36
rotoreflection 28, 36

S

screw
    axis 11, 15, 104
    groupoid 109, 113
    operator 104, 105
    pattern 108
    principal 111
    secondary 113
setting
    ambiguity of 100, 102, 103, 144
space group 12
    Bravais 98–101, 103, 143
    centred 133, 136, 140, 143
    coupling condition 90
    non-Bravais 104, 107, 143
    symmorphic (see Bravais)
space lattice 74, 100
    see also Lattice
Schwarzenberger, R. L. E. 12, 186
Shubnikov, A. V. 10, 186
Steward, E. G. 10
Stanley, T. E. 10
stereogram 39, 41, 46, 54
symmetry axis 15–18, 161
    principal 17–21
    secondary 17–21
symmetry centre 28
    see also Inversion
symmetry plane
    axial 17–20, 25
    transverse 17–19, 22–25, 161

T

tetrahedral
    group 38, 58
    symmetry 41–43
translation group 88–90
two-dimension
    space group 178–180
    colour space group 181–183

U

unit cell 67, 73–75
  body-centred 75, 80, 81
  edge-centred 94, 95
  end-centred 75, 83, 84
  face-centred 75, 81–83
    see Hexagonal
    see Rhombohedral

V

Venkatarayudu, T. 11, 186

W

Williams, Mary 10
Wilson, S. J. C. 10, 186
Woolfson, M. M. 10, 186